This book is dedicated to:

Vito Colangelo, Sr., and Mary Colangelo

V. J. C.

Alan and Elizabeth Thornton

Walter and Mary Mildred Harmon

P. A. T.

Engineering Aspects of PRODUCT LIABILITY

Vito J. Colangelo

Consulting Metallurgist
Advanced Engineering Division
Benet Laboratories, U. S. Army Research &
*   Development Command*

Peter A. Thornton

Senior Materials Engineer
Benet Laboratories, U. S. Army Research &
*   Development Command*
Adjunct Professor, Engineering Technologies
Hudson Valley Community College

 American Society for Metals
Metals Park, Ohio 44073

05581057

Library of Congress Cataloging in Publication Data

Colangelo, Vito J
  Engineering aspects of product liability.

  Includes bibliographical references and index.
  1. Product safety.   2. Products liability.
3. Forensic engineering.   4. Manufacturing processes.
I. Thornton, Peter A., joint author.   II. Title.
TS175.C64   346.7303'82'024671   80-27704
ISBN 0-87170-103-0

PRINTED IN THE UNITED STATES OF AMERICA

D

620.0045

COL

Preface

The primary objective of this book is to provide the practicing engineer and engineering-science student with information regarding the technical side of product liability. Because of the intimate interaction between the technical and legal aspects of product liability, some legal definition and discussion is necessary. However, we endeavor to concentrate on the scientific, technical, and engineering details of the subject. Hopefully, this approach will illustrate the interface between engineering and the law with regard to product safety and liability.

Practicing engineers engaged in design, materials selection and specification, manufacturing, fabrication, and quality control will find useful information concerning the various pitfalls and problems of these respective production areas, and how to avoid them. Correspondingly, this same type of information should be of value to the overseers of production — namely, management — familiarizing them with the role that engineering does and should play in product liability.

For students of engineering and science, a fundamental background in the subject of product liability is presented in rather broad detail. This basic treatment will prepare students to enter industry armed with at least some knowledge of the problems associated with product liability and its potential consequences for them and their company.

Finally, we subscribe to the principle that education in the complexities of product liability for all technical people, engineers in particular, will provide this country with much-needed assistance in reducing and controlling product hazards, thereby decreasing a company's potential product liability exposure and the inevitable lawsuits that accompany this condition. Such action would increase the company's financial well-being; more important, it would improve the welfare of our nation as a whole by providing products whose safe performance can be taken for granted.

We wish to express our sincere thanks to Ellen Thornton and Betty Ann Melius for their diligent typing of the manuscript and to W. Frey for his artistic drawings, on which the line illustrations in this book are based. We are also indebted to the investigators and writers in both the technical and the legal fields whose work we have cited. Without their efforts, progress in the area of product liability would surely be hindered.

VITO J. COLANGELO

PETER A. THORNTON

Contents

Introduction to Product Liability

> A blind man is not required to see at his peril.
>
> Oliver Wendell Holmes
> Common Law 109 (1881)

WHAT IS PRODUCT LIABILITY?

Product liability has different meanings, depending on the context. In legal terms, product liability describes an action, such as a lawsuit, in which an injured party (the plaintiff) seeks to recover damages for personal injury or loss of property from a seller or manufacturer (the defendant) when it is alleged that the injuries or economic loss resulted from a defective product.[1] However, the term product liability may suggest a different meaning to engineers and manufacturers. From a technical viewpoint, product liability can be thought of as the responsibility of a producer for the proper, safe, reliable performance of his product. Accordingly, product liability is minimized when manufacturers develop their products with safeguards against all predictable kinds of defects, deficiencies, abuse, and misuse. Although this is a noble and desirable goal, equipment breakdowns, quality rejections and scrap losses, and injuries to consumers using (or misusing) products are daily testimonials that it is also a very difficult goal to achieve.

There is other, more dramatic evidence of product failure, such as a catastrophic plane crash where loss of life and property are great — for example, the recent incident involving an American Airlines DC-10 at O'Hare International Airport in Chicago on May 25, 1979.[2] This accident involved a failure of the engine support pylon, resulting in separation of the left engine just after takeoff. The reconstructed sequence of events is illustrated in Fig. 1-1. A few seconds

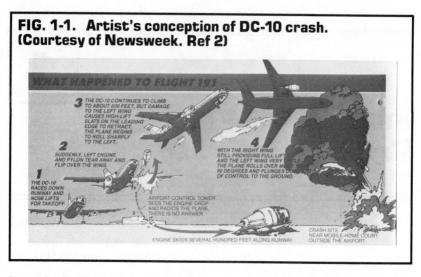

FIG. 1-1. Artist's conception of DC-10 crash. (Courtesy of Newsweek. Ref 2)

later, the huge aircraft nosedived to the ground, killing all on board and two people on the ground, a total of 274 persons. Compounding this tragic loss of life and property, within three days the Federal Aviation Administration grounded the U. S. DC-10 fleet temporarily. Most foreign countries followed this example. The ensuing inspections revealed a number of serious problems in 68 aircraft out of the total fleet of 138. Concomitantly, severe economic losses running in the millions of dollars were suffered by airlines principally utilizing this type of aircraft, such as United, National, and Laker Airways, as a result of lost revenues. The loss of passenger confidence cannot be estimated. Additionally, the technical investigations to determine the cause or causes of the failure and any ensuing litigation will also generate huge costs for the various litigants.

Another pertinent example is the collapse of scaffolding from a partially constructed cooling tower in Willow Island, Virginia, in May 1978.[3] In this incident, the top layer of concrete poured the previous day gave way, causing the supports to pull out and the scaffolding around the entire periphery of the structure to collapse. This "peeling" effect is schematically illustrated in Fig. 1-2. With absolutely nothing to prevent the sequential collapsing, 51 men, the entire crew on the top of the tower, fell 170 feet to their deaths with the scaffolding debris and safety nets tangled around them. In addition to economic losses, product failures can have severe emotional consequences. This disaster was intensified by the fact that the workers were all related or very close friends and came from a small community. Nearly everyone in the surrounding area lost at least one friend or relative; one family tragically lost 10 members!

FIG. 1-2. Illustration of scaffolding collapse. (Courtesy of Newsweek, Ref 3)

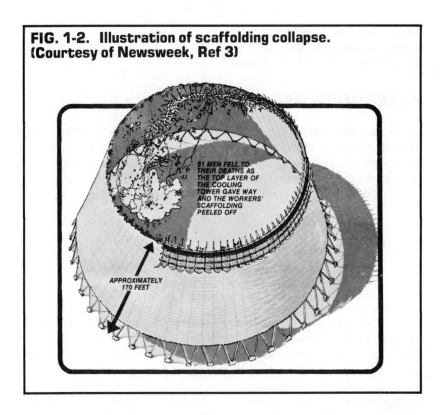

Admittedly, these examples are unusual and are especially shocking because of the extent of personal loss and grief. But they underscore the consequences of defective products or systems and the serious liabilities associated with them, when they involve failures.

Less disastrous, but still very important, are the huge economic losses that accompany minor failures and defective products. Take for example the more familiar product recalls from various manufacturers, particularly the automobile industry because it receives wide attention. The financial losses resulting from replacements, inspection, and servicing of faulty components and systems easily reach into the millions of dollars annually, even without the involvement of any civil litigation. Who do you suppose eventually foots these bills? The costs must be fed back into the system to emerge as a price increase if the company is to remain viable. Additionally, the losses to the United States productivity or GNP resulting from production breakdowns due to defective machinery or equipment, quality rejections due to defects and non-conformity, and finally total scrap losses are virtually impossible to calculate. Furthermore, these categories do not consider the annual costs of product-related injuries and the cost of lost production due to such injuries, which may exceed $6.5 billion.[4]

WHAT DOES PRODUCT LIABILITY ENTAIL?

A realistic product liability prevention program first requires an earnest commitment by management and producers; in other words, the people who make policy in a company. Second, there must be a conscientious attitude on the part of their technical staff responsible for design, development, and fabrication, toward such details as:

> Safe and adequate design
> Proper material specifications
> Process controls
> Quality control, testing, and inspection
> Adequate warnings and labels
> Maintenance and servicing requirements

Once the product is sold and delivered, the old "love-em-and-leave-em" attitude will simply not work in today's society. It has been frequently demonstrated that a manufacturer cannot assume that his product will be used correctly and safely. In fact, a wise producer will not assume even that consumers will use his product solely for the purpose for which it is intended. A popular account of such a situation is the frequently told story about the person who used his rotary lawn mower to trim hedges. As outlandish as this sounds, the tale is not included here for a touch of humor. Rather we mention it because it reflects the unthinking and careless behavior of many consumers regarding the use of a product. Therefore, manufacturers may virtually have to outguess the end users as to the ways and manner of product misuse. This is a primary reason for the current profusion of warning signs, labels, and disclaimers conspicuously posted on power equipment, machinery apparatus, and other potentially hazardous products.

For the engineer and designer, avoidance of product liability involves study, training, experience, and also conscientious planning in the design and manufacture of products. We must continually ask such questions as: Is this design as safe as possible? If not, how can we improve safety at reasonable cost? How might this specific product be misused or abused? How can we prevent such misuse or abuse? How can we warn the consumer of potential hazards, if that is necessary? Are maintenance and inspection procedures which may ultimately affect safety clearly specified and described? Because of the importance of attending to these and other key questions, we believe the role of the technical support staff regarding product liability decisions will continue to expand, especially in industries and companies making high-risk products.

THE CURRENT STATE OF PRODUCT LIABILITY

During the last two decades, product liability litigation has undergone a dramatic rise which has been duly noted in the literature.[5-10] In fact, the increasing number of product liability suits and the problems that accompany this litigation have frequently been labeled a crisis or epidemic. Do we actually have a product liability crisis in this country? As a prelude to more detailed information that will be given in later chapters, let us briefly examine the wealth of pertinent facts presented by Irving[7] and other journalists concerning this aspect of the subject. Fortunately, most of these reports and articles include considerable amounts of data and statistics along with the analyses, to substantiate their warnings regarding product liability.

The Consumer's Side

First, let us examine the consumer's position in the product liability situation. It is appropriate to discuss his plight because he is the most important link in the product cycle. Without the consumer there is obviously no demand, no need for a product and thus no need for a producer. The consumer therefore deserves the utmost consideration. But does he or she actually get it?

In their final report to the President and the Congress in June 1970, the National Commission on Product Safety thoroughly explored and documented the risks, hazards, and consequences associated with numerous consumer products.[11] This Commission was authorized by Public Law 90-146 and appointed by President Lyndon B. Johnson on March 27, 1968. (For the interested reader, portions of this statute, including the duties and powers of the Commission, are presented in Appendix 2.) Alarmingly, the report of the Commission noted that 20 million Americans, roughly 10% of our country's population, are injured annually as a result of consumer-related product accidents. This number represents injuries suffered just in their homes alone! Of this figure, 110,000 are permanently disabled and 30,000 are killed. A very interesting and detailed summary of injuries from certain consumer products is given in Table 1-1. These data

TABLE 1-1. Injuries Associated With Selected Consumer Products Treated In Hospital Emergency Depts. By Quarter and 12 Months Ending 06/30/79

(Source: National Electronic Injury Surveillance System, U.S. Consumer Product Safety Commission/Hazard Identification and Analysis, National Injury Information Clearinghouse)

Report Period, Apr-Jun 1979

Product Descriptions	Number of cases Report period	Number of cases 12 mos to date	Estimated number of cases Report period	Estimated number of cases 12 mos to date	All ages	00-04	05-14	15-24	25-64	65+	Male	Female	Estimated mean severity 12 mos to date
General household appliances													
Cooking ranges, ovens, and related equipment	277	1,209	9,832	36,379	17.1	58.5	9.1	19.3	13.7	10.5	13.4	20.4	100
Irons and ironers, except toys	110	477	2,778	10,998	5.2	34.0	3.8	4.8	2.2	.4	4.7	5.6	68
Refrigerators and freezers	137	705	4,421	20,542	9.6	9.6	6.0	12.8	10.6	5.9	11.8	7.5	33
Space heating, cooling & ventilating appliances													
Chimneys and fireplaces	86	514	2,735	16,391	7.7	65.8	6.9	3.2	1.9	.9	10.2	5.3	59
Fans, electric (exc. stove exhaust fans)	108	514	3,354	12,968	6.1	11.3	4.1	7.2	5.4	6.8	7.1	5.1	22
Heating stoves & space heaters (exc. recreational)	70	413	2,760	15,365	7.2	38.0	5.3	7.6	3.4	4.4	7.9	6.4	142
Pipes (hot water, steam and unspecified)	221	893	8,323	29,474	13.8	13.2	23.5	15.8	11.6	3.1	23.5	4.6	29
Radiators, home and unspecified	112	566	3,260	12,636	5.9	26.5	7.7	6.1	2.5	2.6	7.9	4.0	57
Housewares													
Cookware, pots & pans	160	659	5,512	18,444	8.7	15.9	7.8	12.7	7.5	2.5	7.0	10.2	83
Cutlery & knives, unpowered	608	2,694	22,749	99,813	46.8	9.5	41.8	70.5	52.7	12.6	57.1	37.0	13
Drinking glasses, glass	450	2,245	15,841	71,566	33.6	23.0	20.5	72.3	31.2	13.2	21.3	45.2	13
General kitchen appl., exc. coffeemakers & teapots	72	345	2,294	10,603	5.0	.9	2.0	8.2	5.9	2.6	4.9	5.0	33
Scissors, manual & unspecified	114	570	4,047	16,567	7.8	7.3	15.4	8.3	5.6	3.8	5.9	9.5	27
Skewers, ice picks & other picks	77	392	2,636	12,184	5.7	5.4	11.3	4.5	4.8	2.7	5.4	6.0	21
Tableware, inc. flatware & accessories	482	2,194	14,497	63,037	29.6	43.9	17.3	50.4	28.3	7.7	18.1	40.4	22
Home communication, entertainment & hobby equip.													
Sounds recording & reproducing equipment (electric)	188	835	7,270	27,685	13.0	52.0	10.7	14.0	8.5	6.9	14.3	11.7	39
Television sets	127	652	4,544	20,555	9.6	52.6	12.4	5.3	4.3	5.2	11.0	8.4	34

Estimated number of product-related injuries per 100,000 population within the contiguous United States which were treated in hospital emergency rooms during the last 12 months ending with this report period.

	C1	C2	C3	C4	C5	C6	C7	C8	C9	C10	C11	C12	C13
Home furnishings and fixtures													
Tables, all types	1,566	6,826	51,624	200,305	94.0	645.5	95.8	43.3	36.1	38.2	111.2	77.2	29
Bathtub & non-glass shower structures	590	2,547	21,487	78,023	36.6	105.2	31.6	24.9	28.3	52.4	35.2	38.0	39
Beds (exc. water beds)	1,164	5,580	37,214	156,871	73.6	402.3	101.2	36.8	24.8	70.2	77.0	70.3	36
Benches	97	399	3,302	12,410	5.8	23.5	11.0	3.4	2.5	3.3	7.1	4.6	26
Carpets & rugs	211	921	8,505	34,770	16.3	22.6	9.0	11.5	15.2	37.3	12.3	20.1	34
Chairs, sofas & sofa beds	1,392	6,139	46,757	185,056	86.8	419.4	73.6	46.7	53.3	89.5	81.1	92.0	32
Desks, storage cabinets, shelves & racks	799	3,467	25,921	103,176	48.4	168.6	74.0	42.0	27.4	22.4	49.6	47.1	25
Electric fixtures, lamps & associated equipment	196	1,047	6,160	31,523	14.8	30.1	14.7	18.7	12.4	6.7	18.5	11.1	59
Furniture, not specified	133	574	4,644	18,646	8.7	14.0	7.0	9.3	9.1	5.4	8.7	8.7	21
Ladders & stools	775	3,091	28,214	96,118	45.1	21.8	17.1	27.9	63.5	58.7	67.3	24.1	43
Mirrors & mirror glass	130	566	5,090	16,256	7.6	13.6	6.6	15.5	5.2	.3	8.2	7.1	16
Plumbing fixtures (sinks & toilets)	179	686	6,811	22,801	10.7	34.5	11.3	9.1	6.1	15.8	10.7	10.7	30
Home workshop apparatus, tools, and attachments													
Automotive tools, accessories & chemicals	143	511	5,356	20,790	9.8	12.0	8.5	16.6	8.7	2.0	16.5	3.3	60
Batteries, all types	126	581	4,421	18,425	8.6	2.2	2.3	17.3	10.3	.8	16.4	1.3	159
Hoists, lifts, jacks & jack stands	369	1,438	13,375	44,670	21.0	11.0	12.8	32.0	24.8	5.4	38.4	4.4	22
Power home workshop saws, inc.	108	533	4,902	18,784	8.8	.9	2.4	16.6	10.5	3.5	16.3	1.7	92
unspecified saws	445	1,867	16,566	69,717	32.7	3.6	11.0	41.1	42.5	31.9	63.8	3.2	71
Power home workshop tools, exc. saws	112	500	4,962	20,099	9.4	.2	3.5	16.7	11.1	4.2	17.9	1.3	29
Screwdrivers	93	396	3,693	13,194	6.2	3.2	3.4	9.0	7.6	1.8	11.0	1.6	17
Wires/cords, not specified	158	513	6,108	18,636	8.7	4.4	11.2	9.9	8.7	5.2	12.5	5.1	23
Workshop manual tools or accessories	221	785	8,843	29,831	14.0	5.2	6.8	24.7	15.8	5.3	25.3	3.2	27
Home and family maintenance products													
Bleaches & dyes, cleaning agents & caustic cpds.	370	1,557	13,457	45,951	21.6	106.9	14.4	19.0	15.1	5.1	22.4	20.7	95
Liquid fuels	277	918	9,878	28,227	13.2	38.7	14.2	17.6	9.2	2.7	22.4	4.5	277
Miscellaneous household chemicals	126	735	4,107	23,101	10.8	62.5	5.8	13.3	5.4	1.7	13.1	8.7	137
Paints, solvents & lubricants	200	681	6,539	19,723	9.3	41.3	5.8	10.2	6.8	1.2	12.0	6.6	82
Packaging and containers for household products													
Cans (inc. fuel storage tanks)	637	2,666	21,282	80,777	37.9	77.1	48.0	40.4	31.5	12.5	40.8	35.1	21
Glass bottles & jars	961	4,131	36,634	123,703	58.0	75.0	87.7	93.7	40.4	8.4	71.2	45.3	18
Sports and recreational equipment													
Baseball, activity & related equipment	7,642	14,396	249,247	416,511	195.4	41.5	334.0	378.4	135.9	1.4	290.6	105.0	22
Basketball, activity & related equipment	2,912	12,329	85,377	381,324	178.9	4.6	236.0	562.6	67.6	.2	299.0	64.7	17
Bicycles & accessories	5,688	18,094	188,393	489,819	229.8	256.3	859.3	199.4	43.0	13.7	313.6	150.0	34
Snow skiing & related equipment	25	1,704	973	56,294	26.4	.2	31.5	67.6	17.5	.8	31.4	21.6	22

(continued on next page)

Table 1-1 (continued)

Report Period, Apr–Jun 1979

Product Descriptions	Number of cases — Report period	Number of cases — 12 mos to date	Estimated number of cases — Report period	Estimated number of cases — 12 mos to date	By age — All ages	00-04	05-14	15-24	25-64	65+	By sex — Male	Female	Estimated mean severity — 12 mos to date
Sports and recreational equipment (continued)													
Bowling, activity & related equipment	122	610	4,001	18,344	8.6	4.5	7.2	12.7	9.5	2.4	8.1	9.1	16
Exercise equipment	295	959	10,608	31,473	14.8	11.1	23.2	37.3	5.9	1.3	22.6	7.4	20
Fishing equipment	500	1,388	20,656	50,235	23.6	12.0	53.6	21.5	17.8	9.6	37.9	9.8	21
Football, activity & related equipment	667	15,048	17,903	407,923	191.4	4.3	448.7	523.0	28.6	—	371.5	20.4	30
Golf equipment (inc. golf carts)	238	647	7,776	19,691	9.2	9.2	14.8	6.8	8.6	7.0	14.4	4.3	36
Guns, air & spring, & guns not specified	249	1,025	10,744	37,728	17.7	5.0	38.5	29.1	10.1	4.0	29.4	6.6	74
Gymnastics & associated equipment	354	1,552	12,440	50,333	23.6	4.0	84.3	40.1	1.8	—	17.6	29.2	18
Hockey (field & ice) & related equipment	237	1,838	4,821	44,946	21.1	—	33.6	65.5	6.1	.1	37.2	5.7	22
Ice skating	68	884	1,883	28,578	13.4	.5	36.1	17.0	7.7	.4	13.0	13.8	23
Mopeds, minibikes & other such vehicles	353	890	14,443	30,494	14.3	3.9	34.3	26.5	6.2	1.0	23.9	5.2	42
Playground equipment	1,844	5,431	63,260	161,952	76.0	312.8	268.1	16.3	4.3	1.2	81.3	70.6	28
Skateboards	493	2,525	15,164	63,316	29.7	12.7	109.7	40.7	3.7	.5	45.3	14.9	21
Snowmobiles (inc. apparel & protective gear)	4	434	149	21,427	10.1	2.4	9.1	23.0	8.7	.1	16.3	4.1	35
Soccer, activity & related equipment	634	2,934	17,818	74,307	34.9	.6	85.2	83.5	8.7	—	53.9	16.7	17
Swimming, swimming pools & related equipment	587	2,533	21,871	69,620	32.7	38.1	81.4	53.4	11.0	.6	43.1	22.7	84
Tennis, badminton & squash, activity & equipment	762	2,667	22,886	75,681	35.5	2.1	17.1	59.0	46.0	3.1	50.5	21.2	18
Toboggans, sleds, snow discs & snow tubing	4	1,171	588	51,003	23.9	10.3	80.7	30.4	6.7	2.5	32.3	15.6	33
Track & field activities, apparel & related equipment	307	533	13,296	22,387	10.5	.3	19.9	31.6	2.2	.5	13.6	7.6	23
Trampolines	93	388	3,508	14,427	6.8	6.1	24.7	9.2	.5	—	6.6	6.9	20
Volleyball, activity & related equipment	525	2,320	18,154	75,990	35.7	—	41.2	78.6	29.5	1.2	35.6	35.4	14
Water skiing, tubing & surfing & related equipment	160	831	6,458	27,286	12.8	—	5.7	35.3	11.3	—	20.0	6.0	53
Wrestling (organized activity) & related equipment	201	1,495	8,305	61,950	29.1	2.8	49.5	92.5	5.8	.2	53.4	6.0	21
Toys													
Projectile or flying toys (not fuel powered)	358	687	13,371	21,966	10.3	7.0	13.6	30.7	3.5	.1	15.6	5.3	20
Skates & scooters (exc. ice skates & skateboards)	1,037	3,687	34,287	120,028	56.3	10.2	154.1	85.7	26.7	.3	41.1	70.7	20
Toy sports equipment	32	139	1,410	3,541	1.7	6.7	5.0	1.2	.5	—	2.6	.8	24
Tricycles (children's)	115	401	4,176	11,889	5.6	53.5	6.7	.7	.6	.6	6.1	5.0	48
Wagons & other ride-on toys	187	559	6,181	19,715	9.3	84.9	14.0	.9		.3	11.7	7.0	28

Estimated number of product-related injuries per 100,000 population within the contiguous United States which were treated in hospital emergency rooms during the last 12 months ending with this report period

Yard and garden equipment													
Chain saws	212	1,002	9,969	46,893	22.0	.9	5.3	21.7	34.2	11.9	43.2	1.8	29
Garden hoses, nozzles, or sprinklers	90	379	3,535	11,981	5.6	6.2	12.7	7.6	2.6	2.6	6.9	4.4	21
Hand garden tools	235	946	8,759	33,672	15.8	17.4	30.3	15.1	12.2	6.7	21.4	10.5	28
Hatchets & axes	133	610	5,909	28,252	13.3	6.1	17.1	17.7	13.2	4.0	23.7	3.4	24
Lawn mowers, all types	862	2,074	33,615	70,031	32.9	15.5	27.4	29.5	39.2	30.8	52.9	13.6	77
Trimmers & other small power garden tools	121	443	3,575	10,712	5.0	.9	2.2	4.8	7.3	3.3	7.7	2.5	18
Child nursery equipment and supplies													
Baby carriages, walkers & strollers	160	568	5,778	15,994	7.5	89.5	2.3	.5	.6	—	8.8	6.2	36
Cribs, playpens & baby gates	67	358	1,783	9,995	4.7	52.1	.6	.8	1.1	.1	4.9	4.4	28
High chairs & youth chairs	73	334	2,326	10,067	4.7	55.1	.9	.6	.7	.1	4.7	4.6	48
Personal use items													
Jewelry, watches, keys & key rings	185	715	6,473	21,234	10.0	27.9	16.2	13.8	4.8	2.3	7.1	12.7	53
Money, paper & coins, inc. toy money	109	486	3,716	12,326	5.8	50.4	11.0	.3	—	.1	5.9	5.6	102
Outerwear, footwear & clothing accessories	265	1,004	10,724	30,880	14.5	11.1	18.5	21.1	12.3	7.6	11.0	17.8	22
Pencils, pens, & other desk supplies	347	1,402	10,249	41,869	19.6	42.0	62.0	16.3	5.2	.5	23.5	16.0	28
Pins & needles	261	1,475	7,438	36,630	17.2	16.3	26.0	23.3	13.6	7.4	11.0	23.1	23
Razors & shavers, razor blades	336	1,552	9,957	41,721	19.6	21.9	17.2	33.2	17.2	7.8	25.9	13.6	13
Miscellaneous products													
Grocery or shopping carts	139	574	4,189	15,124	7.1	52.4	7.5	2.5	2.3	3.2	7.3	6.9	26
Home structures and construction materials													
Bricks or concrete blocks (not part of structure)	344	1,149	12,505	33,718	15.8	42.2	29.4	16.7	8.1	5.7	22.3	9.6	25
Cabinet or door hardware	81	329	3,514	12,486	5.9	19.1	8.4	6.6	3.3	1.9	6.6	5.1	24
Fences (non-electric or unspecified)	925	2,897	31,446	82,310	38.6	45.5	96.2	49.6	17.1	9.9	57.8	20.4	30
Floors & flooring materials	2,500	8,372	97,007	278,110	130.5	347.8	140.9	113.4	86.5	177.4	131.7	129.0	35
Glass doors, windows & panels	1,823	6,505	62,180	187,387	87.9	94.2	111.8	172.7	58.8	15.1	111.7	65.3	20
Handrails, railings & banisters	167	640	5,102	21,014	9.9	14.9	18.7	13.8	5.4	3.5	11.2	8.6	30
Lumber & paneling, not part of structure	527	2,074	22,575	81,353	38.2	40.8	51.4	42.7	35.9	15.5	57.4	19.9	32
Nails, carpet tacks, screws & thumb tacks	2,434	9,191	80,520	259,710	121.9	110.6	200.6	153.7	101.3	26.5	170.9	75.0	17
Non-electric wire	125	414	6,590	16,297	7.6	2.4	17.4	8.5	5.3	3.5	11.4	4.1	18
Non-glass doors, inc. garage & storm doors	321	1,606	12,214	51,556	24.2	42.9	31.3	25.0	20.2	13.9	28.2	20.4	28
Porches, balconies, open side floors, etc.	446	1,787	15,421	53,312	25.0	74.0	27.3	22.2	18.5	19.5	25.6	24.4	39
Roofs & roofing materials	167	642	7,250	23,626	11.1	2.5	9.3	15.3	13.0	3.9	20.6	2.1	50
Stairs (inc. folding), steps, ramps & landings	5,257	22,429	164,281	625,546	293.5	469.2	214.1	328.3	280.6	290.7	237.4	346.1	32
Walls or panels, interior	252	1,059	8,308	33,967	15.9	46.9	21.5	21.1	8.7	6.8	19.5	12.4	28
Walls, not specified	1,052	3,352	35,962	99,998	46.9	83.2	65.2	71.9	33.1	5.1	59.3	35.1	22
Window sills & frames, & door sills & frames	232	958	7,868	30,922	14.5	56.7	15.3	14.3	8.7	8.5	16.1	13.0	25

were collected by the National Electronic Injury Surveillance System (NEISS), a part of the U.S. Consumer Product Safety Commission. In addition to the detailed categories, this data is also extrapolated to the estimated member of product-related injuries per 100,000 population treated in U.S. hospital emergency rooms. Then the information is presented by age groups. Product-related injury information displayed in this manner can be especially valuable to manufacturers.

As mentioned earlier, the cost of these product-related injuries may well exceed $5.5 billion annually. Another interesting aspect of the economics of consumer-related product injuries, which tends to be overlooked, is the "social cost."[12] These expenses include:

1. Untold millions in unrealized income tax revenues as a consequence of the preventable deaths and injuries.
2. An estimated $446 million in public expenditures, or 1.7 percent of all current expenditures, for health care.
3. 2.1% of total hospital admissions filling 18,782 beds for acute care—a number equal to the annual construction of such beds at a capital expenditure of $600 million.
4. The equivalent of the full-time services of 2,000 family physicians, a number equal to one-fourth the new physicians graduated in 1970.
5. A drop of $100 million in potential retail sales as 100,000 families each year suffer an unexpected reduction of $1,000 or more in their budgets.

If the sheer enormity of the foregoing costs doesn't overwhelm the reader, it certainly must make a lasting impression. No doubt the financial picture has changed significantly through this last decade; however, there is no basis for estimating that the figures, no matter what their actual values, are decreasing. Rather, all signs indicate that they are rising.

The Commission went on to conclude that (1) a considerable number of product-related injuries could have been avoided, if more attention had been paid to reducing product hazards, and (2) the exposure of consumers to unreasonable product hazards is excessive by any standards.

The Cost of Safety

The prevention of product-related injuries is a very desirable aim and one that should be vigorously pursued. But what does this type of prevention cost and who will pay for it? A recent trade journal editorial considered the question of costs involved in reducing product-related injuries.[13] The following excerpts are most appropriate:

> Autos, home appliances, presses, welding equipment and even hair dryers have been under pressure from a whole raft of government regulations and restrictions and product liability suits.
>
> There's just no dodging the product safety issue or thinking it will go away.
>
> People are demanding greater safety and the U.S. Government will see that they get it. There's nothing wrong with these goals. But greater product safety does cost money.
>
> The question is—and, please, I'm not being callous, where is the money going to come from?

This is an interesting question and the answers are far too broad to address in this chapter; however, a thought or two for the reader: Are the costs of improved product safety and, for that matter, product liability costs going to be paid for out of corporate profits? Will these costs be shared by "governments" or will the costs eventually be passed along to consumers in some other manner? The possibilities will be explored in detail in Chapter 10.

The Manufacturer's Plight

Indeed, the preceding statements and statistics are a strong indictment of the consumer products industry as a whole. Is this censure valid? Undoubtedly, some of the criticism directed at industry is warranted. The sheer numbers and different types of product-related injuries speak for themselves. But is the recent increase in product liability suits due primarily to a decrease in the overall quality of our products—in other words, an increase in defective merchandise reaching the marketplace? We believe that the overall quality of consumer products has not decreased significantly, but that other factors, such as increased consumer awareness and advocacy, increased production and demand, and encouragement and pressure from the legal sphere, have contributed to the current product liability situation in this country today.

Recently, *Iron Age* magazine surveyed 843 small companies in the United States.[14] Thirty-two percent of these companies replied that they had been involved in a product liability lawsuit. Since the costs for settlement of these litigations are often extremely high, it is not surprising that 80% of the companies surveyed also reported they carry specific product liability coverage with a regular insurance carrier. In the time period between 1975 and 1977, their premiums for this protection have increased, on the average, between 424% and 657%. As a result, 10% of these businesses have been forced to "self-insure." In many cases this means that a company sets aside a portion of its income to protect against the costs of product liability claims. Similarly, groups of companies in the same business or industry may pool some reserves to spread the risk and thus provide greater protection for the individual members. These "associations" are becoming increasingly popular with small companies engaged in the manufacture of high-risk products.

The survey also inquired into more technical aspects of product safety and the liability situation. When the companies were asked whether they had been forced to redesign any product to make it safer or more foolproof, 30% said yes and 62% said no. Ostensibly, the latter percentage of companies have never encountered any safety-related problems with product design. This is a surprising statistic and one that deserves a little attention here, for it is very unlikely, except in the manufacture of an extremely low-risk product, that no safety-related design changes have ever been performed by these companies. Quite possibly the key issue here is the interpretation of the term "forced" redesign. Was a company compelled to change product design because of consumer complaints, or did they do it of their own volition?

A well-publicized case in point involved a manufacturer of automobile tires.[15] In October 1978, the Firestone Tire and Rubber Company initiated a voluntary recall of certain Steel Belted Radial 500 and Steel Belted Radial TPC passenger car tires. However, this action was initiated *after* the National Highway Traffic Safety Administration (NHTSA) made an initial determination in July 1978 that a defect relating to motor vehicle safety existed in the Steel Belted Radial 500 tires. The damage associated with the failure of a Steel Belted Radial 500 is illustrated in Fig. 1-3. Therefore, a recall agreement was actually negotiated between the NHTSA and Firestone. Regardless of who truly originated the recall, the immensity of the problem is awesome. Under the agreement between Firestone and the federal government, about 7.5 million Radial 500s still on the road may be replaced free of charge with a new Firestone tire of comparable quality.

On the other hand, the response in the *Iron Age* survey to the question about forced redesign is especially interesting from the engineering viewpoint. Indubitably, the potential savings in avoiding future product liability claims is worth the cost of redesign, but we

FIG. 1-3. Failed Firestone Steel Belted Radial 500 tire.

cannot help speculating on the savings in both personal injuries and just plain money that would result from avoiding the need for redesign in the first place by improved laboratory and field testing plus closer scrutiny early in production — that is, attention to complaints, warranty claims, etc. An example of exactly such design modification occurred in approximately 6.7 million Chevrolet automobiles and trucks in the early 1970's.[16] In this instance, the NHTSA informed General Motors Corporation that it was close to declaring the Chevrolet engine mounts (on certain vehicles manufactured in the years 1965-1969) a safety-related design defect under the 1966 National Traffic and Motor Vehicle Safety Act. (The subject mount and a modified version are schematically illustrated in Fig. 1-4.) General Motors responded to this information by instituting a recall. At that time it was estimated that GM would spend between $25 and $30 million to notify Chevrolet owners and make the necessary repairs.

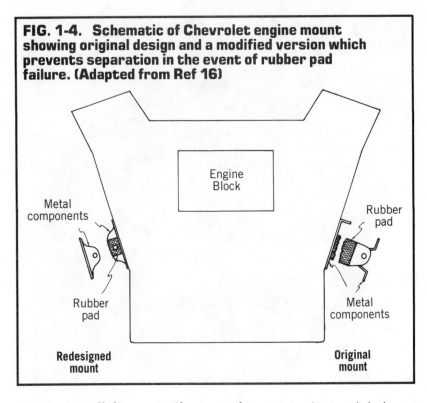

FIG. 1-4. Schematic of Chevrolet engine mount showing original design and a modified version which prevents separation in the event of rubber pad failure. (Adapted from Ref 16)

As we will discuss in Chapter 4, the opportunity to minimize or eliminate product hazards (and thus product liability) often presents itself very early in the production of a component or system. If we can take advantage of such timely circumstances, profits will likely be maximized later on.

Discontinuation of Products

Fourteen percent of the companies in the *Iron Age* survey had been forced to discontinue manufacturing certain "high-risk" products because of the present product liability situation. This is extremely unfortunate for a company, indeed, especially if those specific products were either high on their profit-producing scale or comprised a large portion of a company's market. It is unfortunate for the consumer as well, since if there is a demand for a high-risk product and some companies are forced to discontinue production because of the potential liability risks, a shortage of that particular product will be created. Consequently, the price of that product will undoubtedly increase.

Finally, and most noteworthy, when asked if product liability was a threat to the financial health of their company, over half of the companies replying (54%) answered yes! We could leave the readers to ponder this information and draw their own conclusions about the current state of business with respect to product liability, but one more example is in order which illustrates in capsule form the magnitude of the cost problem to both industry and consumer.

This case involves the Sarlo Power Mower Company in Fort Myers, Florida. The following is a letter from Mr. Arnold Sarlo, the company's president, to Senator Richard Stone in Washington:

Dear Senator Stone:

The Sarlo family has been in the manufacturing business since 1935 (42 years).

Briefly, to introduce ourselves and state our performance record, please read the following:

(a) 42 years of paying taxes and giving many people good-paying jobs.

(b) 42 years of a perfect record of paying our bills.

(c) 42 years of dedicated service to the community too numerous to list here.

(d) 42 years of a spotless record concerning criminal or civil matters.

(e) 42 years of producing and shipping our products to many thousands of satisfied distributors, government agencies, etc.

(f) 42 years of producing a superior product without one product liability suit (minor or major).

We have complied with OSHA standards, insurance standards, been certified by the Outdoor Power Equipment Institute in Washington, been tested by an independent lab and generally complied with any and all regulations to operate as a small manufacturer.

On January 26, 1977 our product liability insurance was cancelled without any reason or explanation (American Fire & Casualty). We also had the same insurance agent for 30 years.

After phoning, begging, talking to our State of Florida representative, State of Florida Insurance Commission, our U.S. Representative Skip Bafalis, the OPEI and so on, no one was able to help us.

Briefly, we went through the worst kind of hell before we managed to get insurance (we hope) through Switzerland (Zurich Co.). The Zurich people are considering us because we have a 42-year perfect record and meet all of their criteria.

Now the worst part; yes, we were happy to find a company to sell us, but compared with our May 11, 1975, year the product liability went up 9,000 pct with a minimum of $45,000 for the first 5,000 power mowers and $9 per unit after 5,000 units. This liability policy covers from zero to $500,000 and has nothing to do with our umbrella policy which is active until Aug. 1, 1977, unless American Fire also terminates it in the near future.

This conceivably means our product liability and umbrella policy could cost (for 10,000 units, our production for the year 5/31/75) $125,000 for a little over a million-dollar volume. This does not include workmen's compensation, other required insurance, inventory taxes, etc.

One thing for sure—we cannot competitively exist.

Sarlo has always had excellent records: you are welcome to examine them. Everything we have in the world is in our company. Do we close and let 42 years of our lifeblood and talent go down the drain—or is there any hope?

We have very few weeks to find an answer (hopefully with your help) or close up and put many loyal employees out of work—and the finest high wheel power mower will disappear from the American market.

Very truly yours,

Arnold L. Sarlo, President

In summary, the Sarlo family business operated for 42 years with a spotless safety record and not one major or minor product liability suit. They complied with OSHA standards, insurance standards, were certified by the Outdoor Power Equipment Institute in Washington, D.C., and had been tested by an independent laboratory. In other words, they were certified, and apparently complied with all regulations governing their industry. On January 26, 1977, their product liability insurance was abruptly cancelled without any reason or explanation, by their carrier of 30 years!

After considerable unsuccessful efforts to obtain help, they managed to find an insurance company that would accept them. However, compared with the year 1975, their liability premium went up 9,000%. Despite their outstanding record, their insurance costs for product liability coverage are now approximately equal to 12$\frac{1}{2}$% of their gross annual sales. Conceivably this additional cost can cut their competitiveness to the point where it is impossible to stay in business.* Unfortunately, this example is not a rare circumstance. Escalation of product liability suits and steeply rising insurance costs are taking their toll on small businesses everywhere in America.

Obviously, then, the role and function of the attorney are not absolutely clear, but are dependent on the circumstances in each case and on his interpretation of the law. It may well be argued that he has an obligation to bring his client's case to the courts for adjudication.

---

*We are pleased to report that the Sarlo Power Mower Company is still operating successfully. Also, as of this writing, they did not have any product liability claims against them. However, because of the astronomical costs of product liability insurance, they are operating on a basis which Mr. Sarlo refers to as "self-insured."

Involvement of the Legal Profession

Certainly responsibility for the current state of product liability does not rest solely with the manufacturers of consumer products. It has been held that a portion of the problem may be attributable to certain parties that stand to gain from product liability suits and litigation, such as attorneys. The inference is that because many legal firms and lawyers work on a contingency fee basis, they may influence injured or irate consumers to engage in unscrupulous civil actions and product liability litigation that is poorly justified. Very often such suits do not reach the courts either because the defendant companies involved wish to avoid undue publicity, or because they are a nuisance and the cost of litigation exceeds the cost of settlement. Consequently, the suits are settled between the litigants. Under these circumstances, the attorney for the plaintiff receives his fee without arguing the merits of the case in court. Just as engineering experts in product liability litigations are sometimes referred to as "hired guns," attorneys that pursue or promote product liability suits are occasionally called "ambulance chasers."

While in some cases an attorney's actions may be questionable, the vast majority of the legal profession are able and ethical, serving simply to present their respective clients' cases to the jury or court. This is where the ultimate decision regarding culpability will be made. At the present time there is no other body to make a decision on the facts in a case.

To illustrate the complexities involved in determining the validity of a plaintiff's case, take the following example. As reported by the Society of Manufacturing Engineers (SME),[17] a foreman in a California stamping plant disconnected the safety devices on a press while making die repairs during a pressing operation. In the course of these repairs, he lost several fingers. One lawyer he contacted told him he had no grounds for suit because he had disconnected the safety system. However, a second attorney informed him that he did have a case and helped win an award of $100,000. Furthermore, this plaintiff, reportedly, is also suing the first attorney for malpractice.

Judicial Involvement

Even our courts are participating to some degree in the intensification of product liability actions. In a product liability briefing session sponsored by the American Management Association in New York, Michel Coccia, a noted defense attorney, who by the way also holds an engineering degree, stated:

Our courts are forever being conditioned either by their own practice or by the plaintiff's bar. They are forever liberalizing basic definitions, twisting rules of evidence and generally favoring the application of principles which will give all the discretionary assistance possible to the injured plaintiff.

The Role of the Media

The news media present reasonably accurate, detailed, and timely reports on many incidents that involve product failures or hazards. However, they may also be guilty of indirectly promoting product liability suits when they devote an inordinate amount of publicity to a catastrophic accident, or headline a large monetary award to a successful plaintiff, simply because the details (often gruesome) of these incidents make good copy. In fact, jury selection is made more difficult in some cases because of the news coverage. Certainly, since only large awards are newsworthy, this accentuated coverage produces a bias in the public's mind and perhaps heightens their expectations.

Engineering Involvement

Regardless of who shares the blame for the product liability situation in this country today, the engineering community must take greater notice of the problem and become more involved if the crisis is to be minimized. Engineers and many other technical people for that matter are periodically obliged to deal with some aspect of product liability because of their academic education, on-the-job training or experience, and position of responsibility within a company. They may be called upon to conduct pre-production safety reviews or investigate a product failure. In addition, they may testify in behalf of their company in a product liability lawsuit.

At present the graduating engineer has very little or no background at all in product liability. A few courses in engineering curricula today are legally oriented, but usually from a management standpoint, and these courses are not generally encouraged by the engineering departments. Unfortunately, this shortcoming in academic development is beginning to surface on the job, especially in small companies where engineers are promptly guided into positions of responsibility involving all facets of production. Design and manufacturing decisions may be made without proper regard for avoiding product hazards. The product liability risks far outweigh the efforts required in producing engineers, scientists, and technicians with some degree of product liability training.

This is a situation wisely recognized by most professional societies today. Many societies offer training to their members in the form of courses (both classroom and correspondence), conferences, and seminars dealing with various aspects of product safety, failure analysis, and product liability.

Personal Liability

Recently, more emphasis has been placed on bringing individuals responsible for unsafe or hazardous designs and products into product liability suits. This reflects directly on the engineering profession because of the large number of engineers and technical personnel that are involved in the critical stages of production such as design, processing, testing, etc. Traditionally, companies or corporations have been sued in product liability suits involving product-related injuries. However, this practice may be undergoing a gradual transition placing more importance on individual liability, especially in cases involving gross negligence on the part of a technical person in a highly responsible position. Potentially, this may involve designers (particularly of critical parts or systems), design reviewers or approvers (drawings or blueprints), quality control personnel (particularly managers who sign certifications for products), right on up the ladder to upper management levels.

Job Security

Finally, the engineering disciplines are subject to the same concerns regarding job security as any other category of employee in a company. We have cited statistics and examples showing financial hardships imposed on small companies by product liability suits and product liability insurance premiums. Obviously, whenever a product or product line is dropped because it increases the company's liability, jobs are jeopardized. When a company can no longer operate profitably because of the costs associated with product liability and goes out of business, its employees join the ranks of the unemployed; then everyone loses.

Similarly, another aspect of job security involves "whistleblowing" by members of the engineering or technical staff of a company. Technically trained personnel are often in a position to spot what they believe to be product-related safety hazards. If the circumstances show the product to be hazardous, which incidentally would increase product liability exposure, a wise company generally takes action to eliminate the hazard immediately. Occasionally, however, a reported safety hazard will go unheeded, for whatever reasons. This situation presents a dilemma to the technical employee. Does he ignore the fact that a hazard exists in order to avoid being labeled a troublemaker and possibly "let go"? Or does he "blow the whistle" on the company and suffer the consequences? A comparable predicament will be discussed in Chapter 3, where we deal with the subject of ethics and integrity during product liability investigations.

THE GOVERNMENT AND PRODUCT LIABILITY

Federal Involvement

The federal government has enacted a substantial amount of legislation relating to the protection of consumers against hazardous products. The following acts are especially pertinent to the current product liability environment. The engineering disciplines involved in manufacturing consumer products should be aware of their existence and familiar with their contents:

1. Food, Drug and Cosmetics Act — 1938
2. Flammable Fabrics Act — 1953
3. Refrigerator Safety Act — 1956
4. Federal Hazardous Substances Act — 1960
5. National Traffic and Motor Vehicle Safety Act — 1966
6. Radiation Control for Health and Safety Act — 1968
7. Poison Prevention Packaging Act — 1970
8. Consumer Product Safety Act — 1972
9. Magnuson-Moss Warranty Act — 1975

For the interested reader, the important details of legislation pertaining to engineering and design are presented in Appendixes 1-3.

The government is also involved in product testing. Although there are over 130 individual government laboratories representing 7 departments and 8 major agencies, only a few agencies actually conduct tests on consumer products to determine compliance with regulatory or purchasing requirements of government. These include the General Services Administration, National Bureau of Standards, Food and Drug Administration, and other smaller agencies.

State Laws

State control of product hazards is limited by jurisdictional boundaries (state lines). State statutes often display little concern for citizens of other states, sometimes specifically denying them the protection afforded local citizens. For example, the West Virginia Code prohibits manufacture, sale, or use of fireworks within the state's boundaries, but exempts manufacture and sale out of state. A state can regulate only the local distribution of products, not the manufacture of products designed and fabricated outside its boundaries.[18] As a result, state regulation of design or composition of a product is rare. Where states seek to regulate design, fabrication, or other elements that could affect safety hazards, they commonly rely on industry or federal standards.

The state of Utah passed a product liability reform act on May 10, 1977, specifically regarding product liability. This act is a first of its kind and the details are worth mentioning because other states will be looking closely at its content. It includes a statute of limitations worded as follows:

> No action shall be brought for the recovery of damages for personal injury, death, or damage to property more than six years after the date of initial purchase for use or consumption, or ten years after the date of manufacture of a product, where the action is based upon or arises out of any of the following:
> 1. Breach of implied warranty
> 2. Defects in design, inspection, testing or manufacture
> 3. Failure to warn
> 4. Failure to properly instruct in the use of a product
> 5. Any other alleged defect or failure of whatsoever kind or nature in relation to a product.

Another section of this act addresses punitive damages as follows:

> No dollar amount shall be specified in the complaint filed in a product liability action against a manufacturer, wholesaler or retailer. The complaint shall merely petition for such damages as are reasonable in the case.

Finally, a section addresses a significant problem encountered by many manufacturers — namely, the alteration or modification of their product after it has been sold:

> No manufacturer can be held liable if the alteration is a substantial contributing cause of injury.

The implications and effects of this legislation on the product liability situation will be discussed further in Chapter 10.

PRODUCT LIABILITY ABROAD

According to the National Commission on Product Safety, governmental action in the United States to protect consumers against hazardous products has been less comprehensive in scope and less extensive in delegated authority than programs in the United Kingdom, Canada, West Germany, Sweden, and Japan. Product safety legislation in these nations gives the national government much broader authority to regulate hazards in the design and manufacture of consumer products than exists in this country.

Traditionally, U.S. producers have exported large amounts of consumer goods and equipment such as machine tools, industrial machinery, agricultural implements and equipment, electronic devices, etc. It may come as a surprise to many readers that some of the legislation in these foreign countries has been applied to products manufactured by American companies on their soil and also to U.S. products imported for retail sales. In light of this potential product liability situation, we feel it will be beneficial to introduce some aspects of their legislation so that readers who are unaware of foreign developments may see in general ways what the atmosphere is in several foreign countries. When analyzing the safety aspects of a particular product, U.S. companies engaged in exporting consumer products to these countries should consider the respective legislative requirements and authority. Often these requirements create contractual problems and may impose restrictions and delays on the importation of the product until the governing authority is satisfied.

United Kingdom

The British Parliament passed a Consumer Protection Act in 1961.[19] Briefly, this act grants the Home Secretary broad authority to impose regulations which are deemed expedient to prevent or reduce risk of death or personal injury from any class of consumer goods or components. Such requirements may relate to the composition, contents, design, construction, finish, or packaging of goods or any component part.

Canada

The Hazardous Products Act was instituted in Canada in 1969.[20] Under Section 7 of this act, the Governor in Council is empowered to issue regulations on advertising, sale, or importation of any hazardous product referenced by a schedule in the act. U.S. manufacturers should take note of this last aspect of the act which directly affects products being imported into Canada. The schedule is divided into two parts, covering those products which are effectively banned from commerce and those which must meet government regulations.

Either schedule may be amended by an administrative order of the Department of Consumer and Corporate Affairs to add or delete any product designed for household, garden, or personal use; in sports or recreational activities, for use by children; and for other purposes. Orders take effect immediately but are subject to review at the request of any manufacturer.

The punchline—and U.S. producers should be fully aware of this—is that the act is enforced through criminal sanctions. A violator is punishable by a fine, imprisonment for 6 months, or both![21]

West Germany

The Law on Appliances (*Gesetz über technische Arbeitsmittel*) of June 1968 gave the government of West Germany broad authority to set technical standards and otherwise regulate the safety of consumer products.[22] It applies to household appliances, sporting goods, do-it-yourself products, and toys. Section 3 prohibits manufacturers or importers from introducing into trade the affected products unless they are manufactured as follows:

> According to generally accepted technical rules as well as the provisions of labor safety and accident prevention, in such a manner that the user or a third person using them as intended, is protected against dangers to life and health to the extent the nature of the use permits.

This legislation takes state-of-the-art into consideration, as well as regulatory standards in manufacturing products.

In summary, it is clear that several nations have established a legal basis for regulating unreasonable risks in consumer products. Furthermore, in some instances, product liability is treated as a criminal offense rather than a civil offense as in this country. These stern measures are certainly worth careful consideration, not only by management, but by both the technical staff (engineers especially) and the manufacturing facility of any company producing goods for consumption or use in the respective foreign countries.

REFERENCES

1. Weinstein, A.S., Twerski, A.D., Piehler, H.R., and Donaher, W.A., *Products Liability and the Reasonably Safe Product*, John Wiley and Sons, New York, 1978, p. 4.
2. "How Safe?," *Newsweek*, June 11, 1979, p. 34.

3. "The Deadly Scaffold," *Newsweek*, May 8, 1978, p. 42.
4. Wulff, J.J., "Selected Data Bearing on the Use of Resources for a Product Safety Program," National Commission on Product Safety, Apr. 1970, p. 20.
5. Shankula, R.D., "An Insurance Engineer Looks at Product Liability," *Mat. Res. and Stds.*, Vol. 9, No. 12, 1969, p. 9.
6. Wise, C.E., "Carnage in the Courtroom," *Machine Design*, May 4, 1972, p. 20.
7. Irving, R.R., "The Awful Truth About Products Liability," *Iron Age*, Sept. 27, 1976.
8. Irving, R.R., "Our National Products Liability Crisis and Why You Are Part of It," *Iron Age*, Aug. 1, 1977, p. 81.
9. Powell, G.W., and Mahmoud, S., "An Engineer's Overview of Product Liability," *Metallography in Failure Analysis*, Plenum Press, New York, 1978, p. 287.
10. "The Devils in the Product Liability Laws," *Business Week*, Feb. 12, 1979, p. 71.
11. Final Report, National Commission on Product Safety, Washington, D.C., June 30, 1970.
12. Ibid., p. 68.
13. Beaudet, G., "Beyond the Grounding of the DC-10's, Editorial, *Iron Age*, June 18, 1979, p. 7.
14. Irving, Ref 8, p. 88.
15. "The Big Firestone 500 Recall," *Consumer Reports*, Apr. 1979, p. 199.
16. "Chevrolet's Failing Engine Mounts," *Consumer Reports*, Feb. 1972, p. 118.
17. Irving, Ref 8, p. 131.
18. Ref 11, p. 81.
19. United Kingdom, Consumer Protection Act (1961), 9 and 10, Elizabeth 2.
20. Canada, Hazardous Products Act, 17 and 18, Elizabeth 2.
21. Ibid., Section 3.
22. Federal Republic of Germany, "*Gesetz über technische Arbeitsmittel*" (Law on Appliances), Bundesgesetzblatt, June 28, 1969, Part 1, No. 42, Z1997A.

Historical Review of Product Liability

To understand the Present,
it is necessary to appreciate the Past.

EARLY DEVELOPMENTS IN PRODUCT LIABILITY

In the study of any complex and controversial subject, it is usually valuable and often interesting to examine the historical events that contributed to its development. This is certainly true for product liability and most appropriate, because it involves the participation and interaction of so many different disciplines. Primarily these include design, engineering, manufacturing, processing, quality control, marketing, legal and judicial principles, insurance, and, of course, consumers. Therefore, we will attempt to present an abridged overview of the history of product liability so that the reader with a technical background may develop a better understanding of the evolution of product safety and product liability and the decisions responsible for this development.

Product liability goes back historically much further than perhaps we realize. For example, in about 1250 B.C. King Hattushish of the Hittites sent a message to King Ramses II of Egypt concerning the delivery of a defective dagger.[1] His letter stated:

> As to the good iron in my sealed house in Kissuwadna, it is a bad time to make iron, but I have written ordering them to make good iron. So far they have not finished it. When they finish it, I will send it to you.

Perhaps we are exercising a little historic license with this example, but the point is that even in those early days "consumers" were rejecting defective products and producers apparently were responsible for making good on them. Furthermore, it is conceivable that "judicial" punishments handed down in cases like that in the above example could be extremely harsh as compared to modern-day punitive measures.

The Early Years

In the centuries that followed the birth of Christ, the rights of common people were virtually nonexistent. Laws were usually enacted by the ruling class and served to benefit them almost exclusively. However, in 1215 A.D., the Magna Carta affirmed the rights and privileges of all classes of free men in England. Although this charter did not deal specifically with any form of product liability as we know it, very likely its articles formed the basis for special laws passed in that country in the 13th century which imposed criminal penalties on tradespeople such as brewers, butchers, cooks, etc., who sold contaminated food and drink.[2] These statutes were very apropos because foodstuffs spoiled quickly and posed a serious threat to consumers. Thus, in a small but nevertheless positive manner, the liability and responsibility of producers was established.

Industrialization. Later, during that country's Industrial Revolution, in the late 18th and early 19th centuries, many new, intricate products were invented and manufactured for public use. In addition, completely new industries were born, such as electrical, photographic, food canning, gas lighting, rubber, and petroleum. This increase in production and corresponding increase in consumption necessarily promoted the issue regarding responsibility of the manufacturer and the retailer or seller. At that time, two points of product liability were recognized:

1. The relationship between the parties entering a contract, specifically either by an express warranty or by an implied warranty.
2. Common law decisions on negligence, involving the perpetration of a tort.*

Privity Rule. In the 1800's, the anticipation of injuries involving a consumer product extended no further than the direct purchaser. Courts were inclined to restrict the liability of the seller or manufacturer as narrowly as possible. The case of Winterbottom vs. Wright,[3] decided in 1842 in England, enunciated the requirement or rule of Privity of Contract between the parties of a negligence suit.

---

*A tort is defined as any wrongful act, damage or injury done willfully or negligently (not in the context of strict liability).

The plaintiff, Mr. Winterbottom, was injured because of defects in a horse-drawn mail coach which the defendant had contracted with the Postmaster-General to keep in repair. The decision, in favor of the defendant, maintained that the manufacturer could foresee injury only to a consumer who had bought his product directly. Apparently, the judges in this era were concerned that a decision for the plaintiff might "open the door" for a multitude of similar cases. This concern, rather than a desire to protect commercial or industrial defendants, is the reason most generally given in the opinions for enforcing the privity rule. Further complicating the privity situation, not only were the potential plaintiffs unknown and possibly unforeseeable by the manufacturer, but in many cases the manufacturers were undiscoverable by the potential plaintiff.

A doctrine such as this was acceptable in an agrarian and tradesman society where people generally purchased tools and equipment for their own use. However, it did not hold up well in the industrialized society that was growing. Extensive mechanization and factory-type production exposed many more people to the harmful effects of a hazardous product. This could occur in the labor-related shopwork where hazardous machinery and equipment, in addition to poor working conditions, were frequently encountered by the laborers and also in the number of defective products reaching the marketplace by virtue of the increase in production due to mechanization. Remember, "mass production" was also a result of the Industrial Revolution. Therefore, as a consequence of industrialization, the privity of contract requirement was subjected to increasingly adverse pressure and gradually changed.

End of Privity. In 1916 in the United States, the case of Mac-Pherson vs. Buick Motor Company[4] ended the era of contractual privity between the manufacturer and the purchaser of his product. Briefly, the defendant in this case was an automobile manufacturer who sold an automobile to a retail dealer. The plaintiff (MacPherson) purchased the car from the dealer. During use by the plaintiff, the car suddenly collapsed, throwing him out and causing injuries. One of the wheels was produced from defective wood and its spokes crumbled into fragments. However, the wheel was not made by the defendant; it was bought from another manufacturer. There was no claim that the defendant knew of the defective condition or willfully concealed it, but there was evidence that the defect could have been discovered by reasonable inspection and that inspection was omitted. The charge was not one of fraud but rather one of negligence. The question to be addressed was whether the defendant owed a duty of care and vigilance to anyone but the immediate purchaser.

Essentially, the court held that a negligence action could be brought against a remote manufacturer of àn automobile with a defectively made wheel that failed, causing injury to the plaintiff.[5] The court went on to say that if a part is fashioned in such a manner that it imperils life and limb, then it is a dangerous product. The consequence of this case was to abolish the special immunity of manufacturers that existed because of the lack of a clearly specified contractual relationship, and to hold them liable for a negligence suit brought by the ultimate user of their product.

Development of Strict Liability. In the years that followed, the principle of Strict Liability in Tort has gradually evolved in the courts.[6] Basically this doctrine means that a manufacturer or producer is responsible for injuries and property damage caused by a defective product. Although the plaintiff must prove that the product was defective and unreasonably dangerous when it was made, the principle of strict liability does away with the privity rule and also eliminates the need to prove negligence on the part of the manufacturer.

As Coccia et al. have emphasized, this principle of strict liability has developed primarily in the interests of consumer safety and protection and rests on the fact that manufacturers are in the best position to provide this protection. Therefore, when manufacturers distribute their products for sale and appeal directly to the consumer through advertising, they become liable and responsible for the safe performance of their products. Courts have consistently ruled that suppliers and manufacturers are responsible for injuries caused by their defective products. The industrial and commercial sector cannot evade the responsibility for such injuries by disclaiming any contractual relationship with a consumer or by claiming that all reasonable care had been exercised during the manufacture of the product.

More Recent Events

The principle of strict liability was reinforced in the 1960 case of Henningsen vs. Bloomfield Motors, Inc.[7] Briefly, Mr. Henningsen purchased an automobile manufactured by Chrysler Corporation, from Bloomfield Motors, Inc. Mrs. Henningsen was injured in an accident allegedly resulting from a steering mechanism failure. She sued both Chrysler and Bloomfield to recover damages for her injuries. Her husband also joined in the suit seeking compensation for his consequential losses. This action was based upon breach of express and implied warranties and also upon negligence. At the trial, the negligence counts were dismissed by the court and the cause was submitted to the jury for determination solely on the issues of implied warranty of merchantability. This was an important development, because negligence can be very difficult to prove and the applicability of "res ipsa loquitur" (the thing speaks for itself) distinctly limited.[8] Therefore, the most direct route to strict liability of remote sellers is through the elimination of the privity in warranty cases. Verdicts were returned against both defendants in favor of the plaintiffs. In other words, the courts held both defendants liable without evidence of negligence and without privity of contract. Remember, Mr. Henningsen alone had executed the contract for the purchase of the automobile, but both he and his wife joined in suing the defendants.

The last example we would like to present in this historical section concerns a further extension of the strict liability doctrine. This is the 1963 landmark case of Greenman vs. Yuba Power Products, Inc.[9] Mr. Greenman sued for damages against the retailer and the manufacturer of a power tool called a Shopsmith that could be used as a saw, drill, and wood lathe. He saw a Shopsmith demonstrated by the retailer and examined a brochure from the manufacturer. Then he received the tool as a Christmas gift from his wife in 1955. In 1957, he purchased the attachments to use the Shopsmith as a lathe for turning a large piece of wood. After working on the piece of wood several times without difficulty, it suddenly flew out of the machine, striking him on the forehead and inflicting serious injuries. A little less than a year later he filed suit against both the retailer and the manufacturer alleging breaches of warranties and negligence.

After a jury trial, the court ruled that there was no evidence that the retailer was negligent or had breached any express warranty, and that the manufacturer was not liable for breach of implied warranty. Therefore the jury received two actions for determination:

1. Against the retailer — breach of implied warranties
2. Against the manufacturer — negligence and breach of express warranties

The jury returned a verdict for the retailer against the plaintiff and for the plaintiff against the manufacturer in the amount of $65,000.

A detailed discussion of this case is given by Noel and Phillips,[10] so it will be sufficient for us to simply point out a few important aspects of the case pertaining to the technical aspects of design and fabrication. The plaintiff introduced substantial evidence that his injuries were caused by defective design and construction of the Shopsmith. His expert witnesses testified that inadequate set screws were used to hold parts of the machine together so that normal vibration caused the tailstock of the lathe to move away from the workpiece, permitting it to fly out of the machine. They also testified that there were alternative, more effective methods of fastening the parts of the lathe together which would have prevented the accident. As we will discuss further in the following section and in Chapter 4, state-of-the-art techniques were not fully utilized in the design and construction of the Shopsmith. This was unfortunate for both Mr. Greenman and Yuba Power Products.

Other Concepts

In addition to strict liability, other theories have recently exhibited increased acceptance in the courts.[11] These concepts are worth mentioning because they are fundamental to the present-day product liability situation:

1. The manufacturer is in a better position than an injured consumer to pay for accidental injuries and damages caused by a defective product.
2. In our highly industrialized and commercialized society, the public interest requires that manufacturers must possess expert knowledge and skill. Therefore, the producer must utilize current or state-of-the-art techniques in his production facility and produce the safest possible product.

RECENT EXPANSIONS OF PRODUCT LIABILITY

The Statistics

In 1963 some 50,000 product liability suits were filed or in progress in the United States.[12] In 1977, Senators J.C. Culver and G. Nelson predicted that the number of product liability suits would reach two million by 1980.[13] Although the data for 1980 are not yet available, Victor E. Schwartz, Chairman of the U.S. Commerce Department's Task Force on Product Liability and Accident Compensation, notes that these estimates are sometimes inaccurate relative to the actual number of cases filed.[14] For example, a figure of one million product liability suits is often cited for the year 1976. Mr.

Schwartz reports that the actual number of suits filed in 1976 is closer to 84,000, a substantially smaller number. Nevertheless, this figure still represents a tremendous amount of costly, time-consuming litigation. Furthermore, we would point out that these cases are the recorded ones that are not settled beforehand.

Correspondingly, the size of court awards from these lawsuits has also increased substantially. Awards of $100,000 to $200,000 are not unusual these days, as compared to an average award of about $11,000 in the early 1960's.[15,16] Most noteworthy in this respect is the infamous product liability litigation surrounding the Ford Pinto. In June 1978, the Association of Trial Lawyers in America reported that there were up to 50 Pinto-related civil suits pending in various courts.[17] Also, at least six suits have been settled out of court by Ford, including three of more than $1 million each. One of the most startling awards in the chronicles of product liability suits was recently made in connection with this automobile when $126 million was awarded by a jury to a quadriplegic survivor of a Pinto rear-end collision.[18] Although this award was subsequently reduced to $5 million, a stigma was cast over the vehicle. As a matter of fact, an Elkhart, Indiana, grand jury indicted the Ford Motor Company on criminal charges on September 13, 1978, in connection with a Pinto automobile crash in which three teenage girls were killed.[19]

Subsequently, in March 1980, a jury found Ford not guilty of the charge of negligent homicide.[20] However, the repercussions of this litigation are numerous. For instance, the company, which had faced a total maximum fine of $30,000 if convicted, reportedly spent $1 million to defend itself. Also, Ford has probably suffered adverse publicity in the face of our present-day depressed economic climate. Furthermore, had Ford lost this particular case, the precedent for "estoppel" or "res judicata" might have been established. In substance, this means "the thing has been decided" and the company could have been barred from litigating further suits lodged against them for the same circumstances.

Recent Legislation

Any legislation which helps to eliminate hazardous products from the marketplace will necessarily reduce accidents and injuries, and, therefore, the number of product liability cases. The most recent comprehensive legislation enacted to reduce the hazards associated with consumer products is the Consumer Product Safety Act of 1972.[21] This act integrated responsibility for administering several earlier legislative acts dealing with consumer safety, including the

Federal Hazardous Substances Act, the Child Protection and Toy Safety Act, the Poison Prevention Packaging Act, the Flammable Fabrics Act, and the Refrigerator Safety Act. The important details of the Consumer Product Safety Act (CPSA) are presented for the interested reader in Appendix 2.

MODERN THEORIES OF PRODUCT LIABILITY

General Principles

In general, before any legal proceedings can be undertaken in a personal injury or liability case involving a consumer product, it must first be determined what legal principles are involved.

From the engineer's standpoint, such clarification can be very helpful to the organization and direction of the technical or scientific investigation. In other words, his findings and subsequent reports should not only be accurate and factual but should also, as far as possible, fundamentally support the legal principles or legal direction his counterpart attorney takes in the case. Then, if the engineer is called upon to give expert testimony, it will complement the other evidence and testimony, thus strengthening the overall case. This approach is appropriate for either side of the litigation, the defense or the plaintiff.

It should be recognized that product liability is not the equivalent of *absolute* liability. The fact that an accident occurred while a product was being used does not automatically place the liability on the product. The legal responsibility of the manufacturer must be established and it must be determined whether the manufacturer has violated his responsibility to the consumer. As Weinstein et al.[22] have stated, these responsibilities are expressed in legal principles which test the conduct of the defendant or the quality of the product. What are these legal theories and principles? The following sections describe in detail the principles utilized in product liability litigation.

Negligence

Negligence from the standpoint of product liability results from the failure of the manufacturer or seller to exercise "reasonable care," thereby exposing the user or consumer to unreasonable risk of harm when the product is being used as intended. The determination of negligence, then, in Weinstein's terms, "tests the conduct of the defendant." The manufacturer's actions are evaluated relative to the technology and level of knowledge, i.e., the state-of-the-art available to him at the time of manufacture. Note that this is based not on what he actually knew but what he should have known. This therefore makes it incumbent upon the manufacturer to be aware of and evaluate new developments in technology which affect product quality and safety. Keeping abreast of the state-of-the art may be accomplished through such means as:

Formal training and education at recognized universities and other educational institutions.

Technical and scientific meetings, seminars, and conferences.

In-house seminars and training.

Membership and participation in professional societies.

Appropriate correspondence courses.

Familiarity with current scientific, business, and industrial periodicals.

Documentation of such endeavors by a company demonstrates a progressive attitude and may prove to be very valuable in possible future product liability litigation involving its products.

The mere adherence to an industry standard does not exculpate a manufacturer from negligence. It has generally been held that standards, voluntary or statutory, are at best the minimum requirements that must be met. If the knowledge is available to produce a safer product at reasonable cost, then the manufacturer alone must decide what is reasonable behavior. If the manufacturer applied reasonable standards of manufacture and quality control, he will generally not be held liable, under the negligence principle, for a defective product which escaped detection. However, it should be noted that lawsuits are decided on a case-by-case basis and it is possible that the same level of care may be held negligent in one court and non-negligent in another. However, since the law is a dynamic, changing process, the manufacturer may determine from the recent body of decisions being generated what constitutes a reasonable standard of conduct.

In essence, we feel that as state-of-the-art advances become available, product manufacturers should expeditiously implement them, particularly if such advances improve the safety of their products.

Breach of Express Warranty

Breach of express warranty occurs when the product does not meet the representations made by the manufacturer with the result that damage or injury occurs. Liability to the producer can accrue if the buyer can prove that his damages resulted from the failure of the product to meet the warranted representatiohs. It is not necessary that such representations be made in writing, as shown by a review of the section of the Uniform Commercial Code (2-313) that deals with express warranties:*

Express Warranties by Affirmation, Promise, Description, Sample

(1) Express warranties by the seller are created as follows:

(a) Any affirmation of fact or promise made by the seller to the buyer which relates to the goods and becomes part of the basis of the bargain creates an express warranty that the goods shall conform to the affirmation or promise.

(b) Any description of the goods which is made part of the basis of the bargain creates an express warranty that the goods shall conform to the description.

(c) Any sample or model which is made part of the basis of the bargain creates an express warranty that the whole of the goods shall conform to the sample or model.

(2) It is not necessary to the creation of an express warranty that the seller use formal words such as "warrant" or "guarantee" or that he have a specific intention to make a warranty, but an affirmation merely of the value of the goods or a statement purporting to be merely the seller's opinion or commendation of the goods does not create a warranty.

Note that the use of formal language or the words "guarantee" or "warrant" is not required for the existence of an express warranty. An express warranty can exist without the incorporation of such language.

Similarly, any language which accompanies the product becomes the potential basis for an express warranty. Good examples of this are found in the catalog descriptions of merchandise offered for sale by mail-order houses. Examples of descriptive language that can form the basis of an express warranty are: heavy duty, commercial grade, skid-proof, 1,000 lbs. carrying capacity, to mention only a few. Photographs, display material, and manuals accompanying a product can also form the basis of an express warranty. It is important to display the product as it should actually be used rather than under circumstances which would extend its use to the point of being hazardous.

Breach of Implied Warranty

An implied warranty is one which is implied by the law rather than being specifically made by the seller. The basis for this rests

---

*Selected articles from the Uniform Commercial Code are included in Appendix 3.

primarily on definitions found in the Uniform Commercial Code, a codification and replacement for several existing acts dealing with sales and commercial transactions. As of 1977, it had been adopted by every state except Louisiana. Two sections of this code, Sections 2-314 and 2-315, largely form the basis for liability suits predicated upon breach of implied warranty. Section 2-314 reads as follows:

Implied Warranty: Merchantability; Usage of Trade

(1) Unless excluded or modified (Section 2-316), a warranty that the goods shall be merchantable is implied in a contract for their sale if the seller is a merchant with respect to goods of that kind. Under this section the serving for value of food or drink to be consumed either on the premises or elsewhere is a sale.

(2) Goods to be merchantable must be at least such as

(a) pass without objection in the trade under the contract description; and

(b) in the case of fungible goods, are of fair average quality within the description; and

(c) are fit for the ordinary purposes for which such goods are used; and

(d) run, within the variations permitted by the agreement, of even kind, quality and quantity within each unit and among all units involved; and

(e) are adequately contained, packaged, and labeled as the agreement may require; and

(f) conform to the promises or affirmations of fact made on the container or label if any.

While the provisions of this section are subject to the interpretations of the court in which the litigation is taking place, and are also subject to the specific legal requirements of the Code, there is, according to Weinstein, a general consensus that this section affords the consumer the same protection as Section 402A of the Second Restatement of Torts (described in the following section on strict liability), i.e., that the product is free of unreasonably dangerous defects. There is still another interesting aspect of implied warranty, that of fitness for particular purpose. Section 2-315 of the Code states:

Implied Warranty: Fitness For Particular Purpose

Where the seller at the time of contracting has reason to know any particular purpose for which the goods that the buyer is relying on the seller's skill or judgment to select or furnish suitable goods, there is unless excluded or modified under the next section an implied warranty that the goods shall be fit for such purpose.

The significance of this section is that a seller may be held liable if he knows the use for which the product is intended and makes an improper recommendation. The implications here are broad, and the transactions which particularly involve engineering fields in which the purchaser relies upon the technical skills and recommendations of his supplier are numerous. For example, small metalworking companies lacking a metallurgical staff often rely entirely upon the advice and recommendations of their steel suppliers in selecting a particular alloy for a component or new application. Should this alloy not perform satisfactorily because of a metallurgical deficiency, it is quite likely that the supplier could be held liable. It is doubtful that many suppliers are aware of the potential risks they may be incurring in making casual recommendations or offering advice based on cursory or secondhand information.

Strict Liability in Tort

Until recently, any personal injury or liability claim was predicated upon the basic theories of negligence or breach of warranty. However, as changes in our methods of marketing and distribution have come about, the law too has changed and a legal concept has evolved designed to protect the remote consumer. As Coccia et al.[23] state, it provides for recovery of damages by an injured user without the need for compliance with the strict requirements of the negligence and warranty law and without twisting other theories.

This legal concept is the theory of Strict Liability in Tort. The law on strict liability has been published in Section 402A of the Restatement of Torts, and the substance of the restatement is as follows:

> One who sells any product in a defective condition unreasonably dangerous to the user or consumer or to his property is subject to liability for physical harm thereby caused to the ultimate user or consumer, or to his property, if
>
>> the seller is engaged in the business of selling such a product, and it is expected to and does reach the user or consumer without substantial change in the condition in which it is sold.
>>
>> This applies even though the seller has exercised all possible care in the preparation and sale of his product, and the user or consumer has not bought the product from or entered into any contractual relation with the seller.

The essence of strict liability is that the manufacturer or seller is subject to liability for damage or injury caused by a defect in a product, if it existed when the product left the manufacturer's control. Furthermore, since publication of the Restatement of Torts in 1965, the courts have extended this protection under Section 402A to include bystanders in addition to the actual user.

Note that the defect must be the proximate cause of the injury. The existence of a defect is not sufficient, since presumably a minor defect could exist in a product with no effect on its function. What may the defects be? Defects may be found to exist which are the fault of manufacturing or production errors. They may also result from improper design or materials selection. The identification and organization of product defects will be discussed in more detail in Chapter 4, in connection with engineering factors that affect product liability.

Nevertheless, an important consideration resulting from the concept of strict liability is that anyone in the chain of distribution, from manufacturer to retailer, may be held liable, provided the product has not been altered. Alteration or modification of a product is a common basis for the defense in product liability litigation. Such modifications may range from the removal of safety guards and protective devices to the extremes of making radical revisions to the basic structure of the product, with subsequent utilization in a manner completely different from the manufacturer's original intent.

Apportionment of Liability

In the ordinary product liability suit for breach of implied warranty, reparation may involve shifting the entire loss backward along the chain of distribution, since the implied warranty is applied to each individual sale of the product. So, if the consumer wins an award for breach of implied warranty from the retailer, the distributor who sold the product to the retailer may be liable for compensating the retailer. Likewise, the manufacturer may be liable to the distributor. Each of these actions may require a suit to establish the apportionment of responsibility. Obviously, this course of legal action is redundant and wasteful of the litigant's time and finances, in addition to plaguing the already overburdened courts.

The need to avoid such improvident litigation was one of the reasons for the adoption of strict liability. In their discussions of Dole vs. Dow Chemical Co.,[24] a case involving third-party action against the employer of the plaintiff's husband, Noel and Phillips[25] comment on the sharing of indemnification by the joint tort-feasors (participants in the wrongdoing). The question is, how should the ultimate responsibility be distributed? There are situations where the facts of a case warrant passing on all responsibility that may be imposed. There are also circumstances where the facts would not warrant passing any of the imposed liability onto a third party. Still other circumstances would justify apportionment of the responsibility between third-party plaintiff and third-party defendants — in essence, an apportionment of indemnification.

Although the law involving apportionment and contribution (sharing responsibility) is not straightforward, the conclusion reached by these legal scholars is that where a third party is found to have been responsible for a part, but not all, of the negligence for which the defendant is adjudged, the responsibility for that part is recoverable by the primary defendant against the third party. To reach that determination, there must necessarily be an apportionment of responsibility for negligence between those parties.

REFERENCES

1. Cited in Herzig, A.J., "Transformation and Hardenability in Steels," Climax Molybdenum Co., Feb. 27–28, 1967.
2. Coccia, M.A., Dondanville, J.W., and Nelson, I.R., *Product Liability — Trends and Implications*, American Management Association, Inc., 1970, p. 10.
3. 10 M. & W. 109, 152 Eng. Rep. 402 (1842). (Noel, D.W., and Phillips, J.S., *Product Liability — Cases and Materials*, American Casebook Series, West Publishing Co., St. Paul, Minn., 1976, p. 98.)
4. Court of Appeals of New York, 1916. 217 N.Y. 382, 111 N.E. 1050.
5. Weinstein, A.S., Twerski, A.D., Piehter, H.R., and Donaher, W.A., *Products Liability and the Reasonably Safe Product*, John Wiley and Sons, New York, 1978, p. 13.
6. Coccia et al., Ref 2, p. 11.
7. Supreme Court of New Jersey, 1960. 32 N.J. 358, 161A. 2d 69.
8. Noel, D.W., and Phillips, J.J., *Products Liability — Cases and Materials*, p. 108.
9. Supreme Court of California, 1963, 59 Cal. 2d 57, 27 Cal. Rptr. 697, 377 P. 2d 897.
10. Noel and Phillips, Ref 8, p. 80.
11. Coccia et al., Ref 2, p. 13.
12. Wise, C.E., "Carnage in the Courtroom," *Machine Design*, May 4, 1972, New York, 1978, p. 13.
13. Congressional Record — Senate, S1752-1753, Jan. 31, 1977.
14. *Business Week*, Feb. 12, 1979, p. 73.
15. Powell, G.N., and Mahmound, S., "An Engineer's Overview of Product Liability," *Metallography in Failure Analysis*, Plenum Press, New York, 1978, p. 287.
16. Shankula, R.E., "An Insurance Engineer Looks at Products Liability," *Mat. Res. and Stds.*, Vol. 9, No. 12, Dec. 1969, p. 9.
17. *Troy Record* (A.P.), Troy, N.Y., Sept. 14, 1978.
18. *Newsweek*, June 26, 1978, p. 56.
19. *Troy Record* (A.P.), Sept. 14, 1978.
20. *The Saratogian – Tri County News* (A.P.), Mar. 14, 1980.

21. Consumer Product Safety Act, P.L. No. 92-573, Oct. 27, 1972, 15 U.S.C., Sections 2051–2081.
22. Weinstein et al., Ref 5, p. 5.
23. Coccia et al., Ref 2, p. 19.
24. Court of Appeals of New York, 1972. 30 N. Y. 2d 143, 331 N.Y.S. 2d 382, 282 N.E. 2d 288.
25. Noel and Phillips, Ref 8, p. 278.

3

The Role of the Engineer in Product Liability Litigation

ENGINEERS AS TECHNICAL EVALUATORS OF FAILURES

Frequently engineers are asked or required to analyze failures associated with product liability cases. Depending on the circumstances, they may have to represent such parties as a supplier of raw materials, an intermediate processor, the manufacturer, an insurance company, or a consumer. Regardless of whom the engineer actually represents, his investigation must be explicit, thorough, and unbiased. The technical evaluation of any failure must uncover and define all the technical factors involved in producing the failure. Furthermore, with regard to liability, the engineer's evaluation should indicate whether or not a defect or defective condition existed in the product prior to the failure. If the investigation reveals that a defect was present in a product before the failure incident, the engineer must then assess the role or contribution of such a condition to the overall failure.

As part of the technical evaluation of a failure, the engineer will also determine whether the product meets any mandatory standards, requirements, or specifications covering it. This examination may include dimensional inspection, mechanical properties, chemical analysis, nondestructive testing methods (radiography, magnetic particle inspection, ultrasonic testing, etc.), or some other physical attributes of the product. Such analysis will determine if the product in question violated the standards or requirements imposed upon it. Comparison with voluntary consensus standards and state-of-the-art practice is also made to determine whether the product was manufactured using up-to-date technology. The product in question is also compared with similar products manufactured by competitors.

Although failures arise from many causes, for mechanical equipment the reasons can be roughly classified into the following three categories:[1]

1. *Design Inadequacies* — for example, sharp corners and radii, abnormal stress raisers, inadequate fasteners, unsuitable material or heat treatment, failure to anticipate conditions of service, and lack of adequate stress analysis.
2. *Processing and Fabrication* — such factors as quench cracks, improper heat treatment, forging or casting defects, nonmetallic inclusions, misalignments, welding flaws, improper machining and grinding, cold working, and improper assembly.
3. *Environmental and Service Deterioration* — including overloads, chemical attack, wear, corrosion, embrittlement, creep, and inadequate or improper inspection and maintenance.

The engineer's investigation, described fully in Chapter 5, will most often encompass several areas of inquiry. Specifically, these will be related to:

> Design of the product
> Materials of construction
> Methods of fabrication and production
> Quality control and inspection procedures
> Service conditions, intended and actual
> Labeling and instructions
> Applicable standards, governmental and industrial
> Current state-of-the-art
> Observations and statements made by witnesses

Factors of design and materials that are likely to affect the investigation are discussed in detail in Chapters 5 and 6. Based upon inputs derived from these areas of investigation, the engineer will attempt to reconstruct the circumstances surrounding the accident and synthesize the cause of the failure. He will then be in a position to determine whether the failure was the result of product deficiency, improper installation or maintenance, or user abuse or misuse. Upon completion of the failure analysis, the engineer is then in a position to decide whether a basis for litigation exists under one of the theories described in Chapter 2. Once this determination has been made, the engineer can assist the attorney in developing the strategy of the litigation, the theories upon which to proceed as well as the anticipated defenses of the opposing litigants.

As an illustrative example, we have created a hypothetical case involving the cast aluminum automotive wheel shown in Fig. 3-1. The wheel as designed by the marketer contains an aluminum rim which is press fit into a steel hub. The aluminum rim was supplied by a subcontractor to the assembler, who produced the assembly and marketed it. Sales literature supplied to the user indicated that the product had been treated with a corrosion guard which sealed the pores and protected against oxidation. The wheel was sold and placed in service in northeastern United States. After three years of use on a passenger vehicle it failed in service, causing injury to the driver. Failure analysis conducted on the subject rim disclosed that it contained shrinkage porosity in excess of recognized standards. In addition, chemical analysis revealed that the composition was not in accordance with ASTM standards in that it contained excessive silicon. The failure analysis revealed that the failure mechanism was corrosion fatigue and that cracking was the result of corrosion occurring between the aluminum rim and the steel hub. The excessive porosity and silicon content decreased the toughness of the alloy and facilitated brittle fracture.

Let us look at the various legal theories again with an eye toward what deficiencies might exist in the product.

(a) Strict Liability. Under the definition of strict liability, the product must have left the seller containing a defect which rendered the product unreasonably dangerous. Note that under this theory the key word is defect. The defect may be one of material, of manufacture, or of design.

Using our hypothetical example, we can see that this wheel could be cited for a materials defect (the excessive silicon), a manufacturing defect (the excessive porosity), or a design defect. While the citation for a materials defect or manufacturing defect may be obvious, a design defect may not be readily apparent. In the subject case, the use of aluminum in contact with steel in a corrosive environment (the salt-laden roads of a northeastern U.S. winter) constitutes a design defect, since the avoidance of galvanically dissimilar metals in a corrosive environment is a fundamental rule of design.

(b) Negligence. In a negligence action, the defendant must have acted negligently, bringing harm to the user.·

FIG. 3-1.   Cast aluminum "Mag" wheel with steel insert which caused galvanic corrosion.

In the case under examination, what might the negligent acts be? The failure to meet the specified chemical and radiographic standards could be construed as negligence on the part of the subcontractor. On the other hand, the failure to inspect the incoming components could be taken as a negligent act on the part of the assembler. In fact, neither exhibited prudent behavior in that a defective product was produced in which the potential for harm was great.

(c) **Breach of Express Warranty.** The product fails to meet the representations of the seller, with damage ensuing.

If we examine the representations of the seller with regard to our aluminum wheel, we find that the sales literature stated that the casting had been treated against corrosion and was suitable for use on passenger cars. It may also be inferred that this included the Northeast, since the wheels were distributed and sold there without exclusion. This representation turned out not to be true: the corrosion treatment prevented general surface-attack pitting but could not prevent the greater problem of galvanic corrosion.

A rigorous, uncompromising, unbiased failure investigation with its attendant conclusions and recommendations is fundamental and extremely important in addressing the liability of a failed product such as the above hypothetical example. This information demonstrates whether or not the product was defective and if it conformed with mandatory standards and requirements. The various techniques and methodology for conducting failure analysis are discussed in Chapter 9.

(d) Breach of Implied Warranty. The product is unfit, unmerchantable, unwholesome, or unsafe.

This theory is not specifically addressed here because of the similarity of this theory and that of strict liability. Both theories deal with the quality of the product but differ in their legal details as discussed in Chapter 2. From the standpoint of the actual product defect, however, there is essentially little difference.

ENGINEERING ETHICS AND RULES OF CONDUCT

The engineer should endeavor to analyze the product objectively with an eye toward discovering whether defects in design or materials actually exist. It is much easier to succumb to the "halo effect" and take the position that the product is excellent and without fault than it is to criticize designs and decisions made by one's co-workers — and by oneself. Too often, the defendant engineer regards the allegations as nothing more than a nuisance created by charlatan consultants and attorneys.

In the event of a real defect, this attitude can place the company in the serious position of being on trial and not having an adequate defense, thereby facing serious financial losses. An admission of the defect by the engineer to the attorney during the pretrial negotiations could improve an otherwise disastrous situation. Willingness to acknowledge that a defect exists benefits the company in another way: early discovery can and should improve a design and reduce the potential exposure to liability suits in the future.

Occasionally, the engineer conducting a failure analysis is subjected to adverse pressure and suggestion from the principals — clients, attorneys, underwriters, or, in the case of an "in-house" investigation, the company's management. These pressures, intended to direct or influence his examination, can be subtle. For example, as the consultant or failure analyst, you may have formed some opinions pertaining to a certain failure that are unfavorable to your client or fiducial party. Then, in a matter-of-fact manner, you are tactfully notified that another expert, perhaps a highly recognized technical person, has said something contrary to your findings. The statement may or may not accurately reflect the thinking of the other expert; nevertheless, the intent of this technique is to undermine your confidence in your own judgment and change your attitude. Other pressures may be overt and strong. For instance, consider the situation where the principals feel so strongly about the outcome of a certain liability case that they virtually issue the engineer an ultimatum to produce a favorable report or be discharged.

Another form of persuasion may be the promise of financial or other rewards for a "job well done" — with the clear implication that this means presenting findings favorable to your client. For the in-house engineer, this may come in the form of a suggested promotion. For the independent engineer, the promise will be increased financial reward in the form of a large bonus. The engineer must of course be adequately compensated for the time he spends on the investigation. This may include time spent on laboratory and test work, inspections, various analyses, photography, etc. The fees are usually set on a per diem basis and as with any other assignment will vary according to the age, experience, and reputation of the consultant. But he should be paid only for his actual *time* and *direct involvement* in a failure analysis or product liability investigation. He should not accept product liability work on a contingency fee basis or become involved in bonus or reward systems for winning cases. Obviously these forms of compensation can bias the technical analyst in his examination and should be strictly avoided!

Unfortunately, unethical circumstances do arise. Therefore, the technical evaluator of failures, especially those involving product liability litigation, must be aware of the psychological persuasions these techniques can have on him, his investigation, and his report. If he becomes aware of slight pressure or subtle suggestion, he should resist, tactfully at first, followed by forceful resistance, if necessary. On the other hand, if overt, heavy pressure is exerted on him and his examination, the engineer must decide whether to be just plain obstinate and accept the consequences or seek employment elsewhere in an environment more compatible with his professional beliefs and attitudes.

THE ENGINEER AND THE ATTORNEY

Under ideal conditions, the engineer or scientist involved in an investigation of product failure will have examined the product soon after the failure. Usually, based upon his examination and inspection of the component, laboratory work, and various documentary evidence, he will have prepared a report. Ideally, he will have maintained control of the hardware and provided for its storage and preservation. In most situations, then, he has entered the case before the attorney and can provide a valuable bridge between past events and those currently occurring.

Most attorneys, even the technically minded, are not familiar with the intricate functions and details of the product they are litigating, unless it is a product to which they have a continuous exposure. It is the role of the engineer then to explain the technology to the attorney. This is not to say that the engineer should treat the attorney like a dimwit. As Kolb[2] points out and any experienced courtroom observer can verify, lawyers have demonstrated many times that they can learn and retain a "startling amount of unfamiliar and complex information."

The engineer is in the position of being the evaluator of the technical merits of the case and of deciding the best way to present those facts and findings which are significant to the outcome. Often he is better qualified than the attorney to know which mechanism or demonstration is available to illustrate a particular point.

He is also in a better position to understand the technical weaknesses in the case, and the competent engineer will communicate this information to the attorney. Only a fool assumes that the opposition will not detect these weaknesses. The wise engineer will assume that the opposition's expert is equally skilled and will discover everything known to him. On the other hand, if the engineer has an obligation to the attorney in the case, so then does the attorney have a similar obligation. Paramount among these is to listen to what his expert is telling him and to be willing to spend the time necessary to gain an adequate understanding of the case.

In the courtroom the expert is in a captive situation, able to respond only to questions that have been asked him. It is through his attorney that he communicates to the jury. Nothing is more frustrating than having an unprepared attorney who stumbles through a series of technical questions or stops short of the critical and fundamental question. The rapport between the attorney and the technical expert should be smooth and cohesive in order to maximize the skills of both.

IN-HOUSE OR OUTSIDE EXPERT?

The question of whether an internal expert or an outside expert should be utilized is one that must be evaluated by the attorney handling a particular case, since both positions have merit.

The outside expert or the independent laboratory is not an employee of a company and is therefore less likely to be regarded by the jury as a captive and nonobjective witness. He is presumed to be able to render a totally objective opinion and to be less sensitive to company pressures.

On the other hand, a number of companies utilize internal experts and in-house facilities to evaluate components, to conduct tests, and to present their findings. These company experts have the entire resources and facilities of the company available to them together with a history of the product. The specialized in-depth knowledge and experience possessed by these experts can be overwhelming when compared to the general expertise of the plaintiff's experts.

However, the possibility also exists that an in-house assessment of a failure will be biased. Similar to the halo effect mentioned in the discussion of ethics, this is due to familiarity and close working relations with that product. We are all familiar with the retort: "What? My [name of product] couldn't have failed like that. It's just not possible!"

THE ENGINEER AS AN EXPERT WITNESS

During the course of litigation, unless the suit is settled out of court or otherwise terminated, the engineer may eventually be called upon to testify as an expert.

Who is the expert witness and how does he differ from the lay witness? The expert witness is one who by virtue of education, training, and experience or a combination of these factors can discuss the operation and function of a particular process or mechanism. He differs from the lay witness in that the lay witness is prohibited from offering an opinion, except in very narrow areas involving human experience and based upon his perception. The expert, on the other hand, can offer his testimony and opinion if the material is sufficiently technical and beyond the understanding of the average lay person — in other words, where the opinion of the expert would provide assistance to the jury or to the court.

Usually the expert witness is an engineer, scientist, or other professional whose specialized education endows him with the knowledge required to understand the matter before the court.

However, the expert need not be a professional. A nonprofessional with years of experience may possess sufficient expertise to testify in a particular area. For example, a skilled auto mechanic may well

be judged competent to testify regarding procedures followed in re-
placing worn brakes.

PRETRIAL PROCEDURES

Subsequent to his technical examination and report, as litigation
progresses, the technical expert may be required to participate in the
"discovery procedures" initiated by opposing counsel. These procedures vary, depending on the jurisdiction. One of these procedures is
the "interrogatory." The engineer, acting under subpoena, must furnish, in writing, the answers to a series of questions. This is usually
conducted at the engineer's place of business in concert with his
counsel. The engineer may use any records, drawings, or other reference material available to him. The procedure followed by most
attorneys is to have the expert answer these questions in the broadest
terms possible, consistent with a truthful reply. However, if the
opposing counsel has adequately prepared his case and is acting with
the assistance of a technical expert, he will reply with a second series
of questions, narrower and more penetrating than the first, until
he is satisfied.

If the litigation is proceeding in a federal court, or in a state using
federal rules, then the pretrial discovery procedure may take the form
of a "deposition." Here the procedure is considerably more formal,
usually taking place in the attorney's office with the company's counsel and a legal stenographer present. The witness is sworn in as in a
normal courtroom trial. The witness must appear with whatever
records and files have been requested and must respond to questions
posed him by opposing counsel, unless instructed not to answer by
his attorney. In the hands of skilled opposition counsel, the experience can be a fatiguing one. The best defense is adequate in-depth
preparation, not only about the issue at hand but about peripheral
matters and the current state-of-the-art.

Obviously, some anticipation of the direction the questioning
will take is required. Preparation in these areas should be undertaken
even in areas that may appear inconsequential. This preparation will
prove of value not only in the pretrial proceedings but in the trial
itself, and may also disclose weaknesses in the theory of the case.

DIRECT EXAMINATION

Insofar as the expert witness is concerned the trial will usually
take the following form. Direct testimony will usually be initiated by
inquiring as to the witness's qualifications. Questions will explore the
extent and nature of the witness's education, training, and experience

as well as his familiarity with the particular component or process at hand. Although some attorneys tend to treat this aspect of the testimony quite casually, it is our opinion that the matter of credentials should be treated as seriously as any other portion of the trial. In this regard, it is far better for an expert to respond to direct questions regarding education, employment-related experience, research, and writing than to answer the general question, "Tell us about your background." In their response to this kind of question many engineers are modest and reticent to a degree that is detrimental. It is important to bring out the details not only of the expert's knowledge of general scientific and engineering principles but also of his experience with similar products and with the state-of-the-art, both currently and at the time the subject equipment was manufactured. In this phase of the trial, a legal procedure known as a *voir dire* may be utilized. The *voir dire* is a preliminary examination to ascertain the competence of the witness. The opposition attorney may ask to examine the witness to establish the extent of qualifications and the validity of the background claimed. The judge too may ask questions to establish the extent of the expert's technical background. As previously stated, the seriousness of this portion of the testimony is not to be minimized. If the trial judge believes that the expert is not competent regarding the issue, he may rule that the expert's testimony is inadmissable and refuse to allow him to testify further.

If the witness is allowed to testify as an expert, a series of questions regarding the details of the accident, the nature of the investigation and tests made, the results of these tests, etc., may then be asked.

When questioned regarding his inspection of the component, the expert should specify when and where he examined it and what his observations were. Here it is important to relate his observations in exact detail — the kind, number, and appearance of any remarkable features and where they were located on the particular component. Photographs taken to illustrate the observations should be presented, as well as the component itself if it is available. If laboratory work was conducted, the witness should disclose what tests were performed and what the results were, together with any additional findings such as radiographic results, chemical analysis, and metallurgical information. It is important to remember in answering these questions that neither the jury nor the judge nor the attorneys are expert in the field about which you are testifying. The purpose of your testimony is to communicate; if you fail to do this, you have failed in your role as an expert. In some cases, it may be necessary to present your technical material twice, first in a precise scientific and technical form and then in an equally correct but greatly simplified version. If necessary, this aspect should be discussed with your attorney, so that the phraseology of the questions is proper and the sequence of questions is thoroughly understood.

The final aspect of direct examination involves the use of hypothetical questions to elicit opinions regarding the safety of the component, the presence of a defect, when the defect came into being, and whether the defect was proximally related to the accident. The questions are predicated on the assumption of facts which have already been introduced into evidence. It is essential to have these facts firmly in evidence and to have them keyed to support the legal theory upon which litigation is based, since without an essential supporting fact, the entire theory may collapse. It is equally important, as Greenstone[3] has stated, that the expert base his reply on probabilities rather than possibilities and be as firm as possible in stating his opinion.

CROSS-EXAMINATION

The function of cross-examination, under the adversary system, is to permit the opposing counsel to inquire into the source of information, the basis for opinion, the specific material upon which it is based, the amount of compensation received, testimony given in other trials or in pretrial depositions, or other matters which would have a logical tendency to rebut inferences drawn as a result of direct examination.

Under cross-examination, the opposing attorney tries to negate the impact of the expert's prior direct testimony. The examiner will usually begin by attempting to disparage the witness's qualifications, concentrating not on his strong points but on the weaknesses in his background. If the witness is a practicing engineer, the cross-examiner may stress his lack of academic credentials and publications. Conversely, if the expert is an academician, then his deficiencies in practical matters may be cited. The witness's lack of previous experience with the component may be the point of inquiry, with the examiner placing considerable emphasis on the differences between this device and others that he might have examined, and little emphasis on the similarities. If the witness is a mechanical engineer, the questions may stress the metallurgical aspects of the product failure. If the witness is a metallurgist, then the inquiry may probe into the mechanical engineering aspects, e.g., design and stress analysis considerations.

Another area which may be explored in an attempt to discredit the witness is that of financial compensation. It will be implied that the expert is being paid to testify — in other words, that he is a "hired gun" if he is an outside expert, or that his job depends on his testimony if he is an internal expert. In either case, the witness's reply should be the same, that he is being compensated for his time and that he is there to present his findings truthfully and honestly.

As regards general attitude and deportment during cross-examination, the following recommendations are valid:

Listen carefully to each and every question. Answer the question asked, not the question you think was asked. Do not volunteer information; answer only what the question specifically requires. Discuss sensitive areas with your attorney prior to your testimony until you are both sure of the ramifications and so that you will not be surprised. If you have any doubt as to the meaning of a question, or if it is ambiguous, ask to have the question repeated. Wait until the question is completed before you begin your answer. On a critical point, watch your attorney to see if he intends to make an objection.

Answer the questions in a firm voice as honestly as possible. If you do not remember a specific fact or event, simply state that you do not recall. Under no circumstances should you guess at a fact or accept one that you do not remember. Answer the questions evenly and in a helpful manner. Do not lose your temper or self-control, even if the cross-examiner is rude or abrasive.

A tactic often used is to get the expert to admit to a specific value, such as the fatigue life of a component, the number of cycles it has experienced, or its weight or another attribute, that has not been studied. The questioning will usually go as follows: "Is the fatigue life greater than a thousand cycles?" "Ten thousand?" "A hundred thousand?" "A million?"

Be cautioned that you should not be taken in by this technique. If you do not know, then you do not know! It is better to suffer the penalty of being uninformed in one area than to be forced into a technical error which may be enlarged and expanded into a major issue.

Another common tactic employed is to pose a series of questions requiring a "yes" or "no" answer. If you can answer the question "yes" or "no," then do so. However, listen to each question carefully and don't be lulled into an erroneous answer by the rhythm of the answers. If you cannot answer the question "yes" or "no," then state that the question cannot be answered completely and accurately with such a reply.

One method of cross-examination favored by attorneys experienced in technical matters is the technique of asking questions on narrower and narrower grounds. The procedure is designed to force the expert into tighter and tighter circles until the answers are more and more specific. The only way to respond to this kind of examination is to have adequate preparation. You will have to know your material better than anyone questioning you. If you have not prepared yourself for the trial, this kind of deliberate, methodical cross-examination will reveal it.

EXHIBITS AND DEMONSTRATIONS

Exhibits such as photographs, slides, charts, drawings, etc., used by the expert in his testimony should be taken or prepared by the expert himself or under his direct supervision, if at all possible. Such a procedure minimizes any confusion as to the content of the exhibit and eliminates problems of continuity which might require an intermediate witness.

The purpose of your visual and oral presentation is to *inform* the court and the jury. Therefore, diagrams and tables should be relatively simple and uncomplicated, suitable for the lay persons receiving the material rather than for your technical peers. Many experts have difficulty rendering material in an uncomplicated manner, as if such a presentation would minimize their scientific status.

Don't be hasty or terse in explaining a visual aid or exhibit. To adequately show a slide, photograph, chart, etc., requires anywhere from one to three minutes. The longer it is shown, the better it will be understood and remembered. Select your exhibits carefully, on the basis of maximum information. It is better to have the jury remember your presentation in terms of a few vivid photographs than fifteen rapid-fire pictures. If you type up tables for a slide, be sure they are clear and legible.

It is our experience that a slide no more than 40 letters or spaces wide is more easily interpreted and absorbed than one containing much more material.

Remember that you are an expert and are presenting a new subject to the jury. Also, the jury may be very heterogeneous: housewives, salesmen, laborers, retailers, etc. Therefore, remember one rule: Be simple. Do not use technical or unfamiliar terms unless you intend to explain them. If you use numbers or data, reinforce their meaning visually and verbally.

If the main points are clearly presented, your findings will be reinforced. This can contribute effectively to the overall litigation.

"Dry runs" with your client, attorney, aides, etc., can be advantageous. You can perfect your timing and delivery and assess whether you are being understood. Ideally, everyone in the courtroom should understand the exhibit and its full meaning.

The same advice holds true for models of a device or mechanism. One particular caution applies to models: if the model is not to scale, it must be so stated. If it is to scale, then the appropriate scale should be stated. Even though they might be easier to make, one must seriously question whether to use models with significant discrepancies from the original. This is true for two reasons: first, there is a good chance that an exhibit will not be allowed if it does not accurately depict the original, and second, even if it is admitted, the opposition may seize upon the weaknesses in the model and attempt to convey to the jury that the entire presentation has been deceitful and fraudulent. This, in fact, could reflect negatively upon the original product. If minor variations in the model are necessary (and they usually are), then it is far better to make your attorney fully aware of the variations and to bring these out in direct examination.

When they are used properly, demonstrations can be dramatically effective in illuminating a critical point, but they hold the seeds of destruction if they backfire. For this reason, demonstrations should be rehearsed over and over until nothing is left to chance.

The authors recall a demonstration in which the defense, placing little credence in the plaintiff's theory that a loose bolt had lodged in the carburetor and apparently spending little time trying to reproduce the conditions, challenged the expert to place the bolt in such a way as to jam the carburetor open. When he did so, as shown in Fig. 3-2, immediately and without difficulty, the defense was devastated and lost the case.

On the other hand, as previously stated, demonstrations can illustrate a point as nothing else can. Take the example shown in Fig. 3-3, in which the flammability of a specimen of polyurethane foam insulation of a particular type is being demonstrated. The demonstration is simple, graphic, and very revealing of the burning characteristics of this material. Similarly, other demonstrations can be used to illustrate or compare the mechanical properties of structural materials. Shown in Fig. 3-4 is a test which demonstrates the impact resistance or toughness of metal or plastic materials.

COURTROOM DEMEANOR

There are several factors which are nontechnical but nevertheless are relevant to the testimony. The expert witness should remember that he is on exhibit as soon as he enters the courtroom. Therefore, dress, speech and general behavior are important. Even

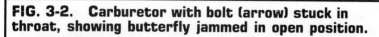

FIG. 3-2. **Carburetor with bolt (arrow) stuck in throat, showing butterfly jammed in open position.**

though these factors have nothing to do with the merits of a case or the competence of a witness, they sometimes weigh heavily in the overall credibility of a witness. Attire should be conservative and not flamboyant. The witness should speak clearly and articulately in answering the questions posed to him, addressing himself both to the jury and to the attorney presenting the questions. Most important of all is sincerity in answering the questions and in demeanor.

Deportment off the witness stand is equally important. Loud, boisterous conversation and actions in the corridors usually are not well received by the jury, particularly if the injuries suffered by the plaintiff are serious.

FIG. 3-3.   Flammability demonstration showing the burning characteristics of foam insulation material. Note burning residue in glass dish.

A final note of caution. If at any time during the litigation the questioning becomes antagonistic or attacking, the expert must strive to remain calm and collected. The popular phrase "Don't lose your cool" truly applies; blowing up in the courtroom, especially in front of a jury, will not help your case. Your part of the litigation will benefit, measurably, if you remain sincere, interested, but imperturbable.

REFERENCES

1. Dolan, T. J., "Analyzing Failures of Metal Components," *Met Engr. Quart.*, Nov. 1972, p. 32.

FIG. 3-4.   Demonstrator testing impact specimen in small impact machine. When hammer is released, it swings down, breaking the specimen. Impact strength is read directly on calibrated dial.

2. Kolb, J., "Engineer as a Witness Must Learn His Lessons," *Prod. Engr.*, Apr. 1969, p. 51.
3. Greenstone, H. E., "What the Attorney Expects of an Expert Witness," Proc. Prod. Liability Prevention Conference, PLP-77E, Aug. 22-24, 1977, p. 177.

4

Engineering Considerations During Design Relative to Product Liability

Quality cannot be inspected into a product. It can only result from prudent design and careful manufacture by responsible people.

THE ENGINEER AS A DESIGNER

No matter what his particular discipline or specialty may be, the engineer frequently must function as the designer of a product. For example, mechanical engineers may be required to plan electrical systems, electrical engineers may find themselves involved in materials specification for motors or solid state circuitry, civil engineers are increasingly involved in sewage and waste disposal, etc. From a strict technical standpoint, this may not be an ideal situation, but in many small companies and manufacturing operations economics necessitate such procedures. In fact, sometimes engineers may find themselves managing or directing all facets of design and construction.

The many qualifications for a successful designer have been thoroughly explored by Nibel and Draper.[1] In general, the engineer or technical person associated with design activities must be versatile, creative, and well informed. In addition to a solid foundation in the basic physical sciences, the designer must possess a comprehensive knowledge of materials and manufacturing processes. Furthermore, he must be familiar with his company's organization and the function of other industries, particularly those with which he will have direct or indirect contact. Finally, the designer should be cognizant of the human element, the physiological and psychological factors involved in both the manufacture and the use of a product. Depending upon the specific product, these factors might include:

Physiological	Psychological
Manual dexterity	Motivation
Hearing	Boredom
Vision	Apathy
Reaction time	Curiosity

Certainly all these abilities help an engineer to function efficiently as a designer. However, among the many qualifications emphasized for effective designers, the ability to perceive potentially hazardous situations or defective conditions is rarely stressed. In fact, during the planning and conceptual stages of design, product liability exposure is usually ranked very low on the scale of important factors influencing product design, if it is considered at all.

Direct consideration of product hazards and product liability potential in the design stage is widely neglected in the product design literature, in training programs in industry, and, unfortunately, in the university classroom. With the exception of a legal-technical course in products liability offered by Carnegie-Mellon University and a course entitled the "Legal Aspects of Engineering" offered by the Department of Chemical Engineering at Ohio State University, we are unaware of any concerted efforts by the technological community in general to deal with product liability at the university level. Special criticism of the academic institutions has been made by the National Academy of Engineering:[2]

> Engineering students are inadequately informed about the consumer needs and value trade-offs. Also, quality and safety are not recognized as the responsibility of each individual engineer but as the responsibility of a few specialists.

We are graduating thousands of engineers annually who have never even heard the term "product liability."

Moreover, courses dealing with design which emphasize nonverbal thinking rather than mathematical or analytical techniques are apparently receiving less and less attention in most engineering schools.[3] As these courses head toward extinction, Furgeson[4] suggests that we will witness an increase in the frequency of costly design errors, especially in advanced engineering systems. His comments on the trend toward abandonment of nonverbal knowledge in today's engineering schools are certainly worthy of consideration. Such an evolution in design practice, if it continues, could have an adverse effect on the product liability situation by increasing the number of design-related failures.

Principal Designer

When he is serving as the primary or sole designer, the engineer is in a position to question the product design and manufacture from a safety standpoint. He can and must anticipate the probable uses and misuses of a particular product and recommend appropriate changes at the earliest possible stage to produce a safe product. Such actions can include relatively minor design changes, major or radical design changes, changes in materials selection and specifications, manufacturing or processing changes, the inclusion of safety features and warnings, and possibly even the ultimate determination — a recommendation to abandon a product altogether if it is hazardous and the risk associated with its use is not justifiable.

Prototype Analysis

During the prototype development and initial testing stages the engineer-designer has the opportunity to observe the function and performance of the intended product in close detail. Scrutiny of the model or prototype component from the standpoint of user safety can provide valuable information with respect to future product liability problems. At this time, the apropos line of questioning should include the following:

What, if any, are the potential hazards associated with the product's intended function?

What hazards may be incurred by improper use of the product?

Can we assess the extent of possible injuries under the above circumstances?

Will a warning statement or label be adequate to prevent such injuries?

Is production of the product feasible or justified in view of attendant risk?

Will a modification or design change eliminate or reduce a potential safety hazard?

Will such a change be cost effective?

Hopefully, the designer is not without assistance at this stage. He may call upon other engineers to act as consultants. Also, the opinions and suggestions of skilled personnel such as technicians, machinists, draftsmen, welders, heat treaters, etc., should be considered in the analysis. Frequently, their close working relationship with a particular process, a manufacturing operation, or a specific material can provide vital insight and important information concerning the hazards a product might present. As we have previously emphasized by several examples in Chapter 1, implementation of the appropriate corrective action, which may well be a design modification, is particularly cost effective when performed in the pre-production stage. Furthermore, such a procedure can result not only in reduced product liability exposure but may also improve product reliability.

For example, a program instituted at Westinghouse Electric Corporation to prevent product liability hazards uses a number of techniques to reduce hazards while simultaneously improving product reliability.[5] These methods include:

The Formal Design Review. All available experience is brought to bear on the design of new products at an early stage in their development, to insure that this experience is used in achieving the optimum design before production starts. The possible interaction of various engineering departments with other operating units of a company during the design stage is depicted in Fig. 4-1.

Failure Utilization. The information gained during a failure investigation is used to determine appropriate corrective action for the design or product involved, explore other product lines for similarities, and add this information to design inputs to prevent similar problems with other new products.

As we have pointed out in Ref 6, although failures are regrettable they are almost inevitable, especially in newly designed products. Therefore, rather than dwell upon them morosely, we must expeditiously recycle the information they provide back into the product's design or processing.

As stated in the preceding section, occasionally it may be prudent to recommend the abandonment of a new product because its design is simply too hazardous. Although this is not an easy decision to make, if circumstances truly warrant such serious action the company's future product liability position will necessarily be improved. Take for example the case of a very successful, highly regarded national manufacturer of power garden equipment who designed and developed a prototype shredder for producing garden compost and mulch. Typically, this sort of machine (pictured in Fig.

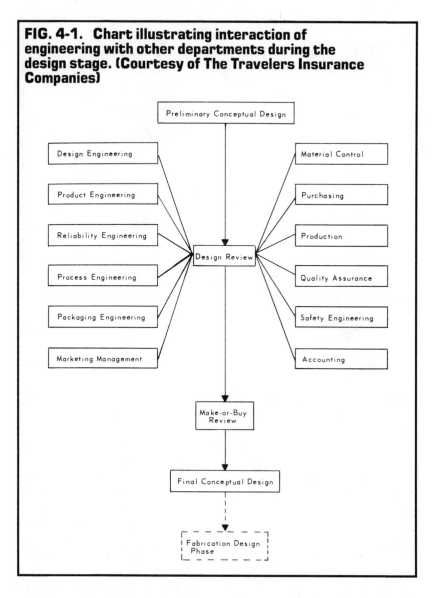

FIG. 4-1. Chart illustrating interaction of engineering with other departments during the design stage. (Courtesy of The Travelers Insurance Companies)

Preliminary Conceptual Design

Design Engineering

Product Engineering

Reliability Engineering

Process Engineering

Packaging Engineering

Marketing Management

Material Control

Purchasing

Production

Quality Assurance

Safety Engineering

Accounting

Design Review

Make-or-Buy Review

Final Conceptual Design

Fabrication Design Phase

4-2) is powered by a gasoline engine and employs a series of cutting blades or knives, in much the same manner as a household garbage disposal unit. The machine is usually capable of chopping and shredding such organic material as brush, leaves, small tree branches, etc., so the resulting product will decompose readily. The company in our example is particularly safety conscious and, as a result, the shredder

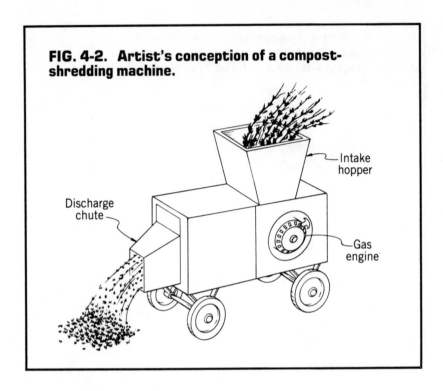

FIG. 4-2. Artist's conception of a compost-shredding machine.

Intake hopper

Discharge chute

Gas engine

design was not lacking in safety features, several of which were quite innovative. In addition, the prototype design included very specific and explicit warnings and labels.

Nevertheless, when the shredder was finally reviewed and analyzed by the full engineering team, it was discovered that under certain conditions, unless the material to be shredded was literally thrown into the machine, the speed of the intake mechanism was so great that it could exceed a person's reaction time to release a branch or similar object. Consequently, someone feeding the machine could possibly have his hand or even his arm drawn into the chopping blades. It is worth noting that their analysis and review addressed the physiological factors affecting the use of a product, with respect to its overall design. The physical or human element of product design is sometimes forgotten in this stage of production. The subject company concluded that there was no simple or practical way to eliminate this hazard, and it was decided that the machine was just too dangerous. Production of the compost shredder was dropped entirely, thereby avoiding the possibility of considerable future product liability litigation for the company, not to mention the possibility of disastrous injuries to a user.

Risk Analysis

Often the design stage is also a suitable time to conduct an analysis of risk. This evaluation may be analytical or statistical in nature. Such a technique has been discussed in detail by McAllister with respect to defects that are presumably discoverable by the user of a product.[7] Certainly, many problems and defective products are discovered by the consumer. The most obvious examples are defects that stimulate the senses — sight, hearing, touch, and smell. Most people are probably familiar with some of the following sensory warnings: vibration in an automobile steering wheel or front end, squealing brakes, noisy bearings in pumps and motors, dripping hoses and connections, smoke or acrid odors from motors or wiring, mild electric shock causing discomfort. There is a certain probability that a user will respond to a warning stimulus and take steps to avoid injury or correct the problem. Consequently, this information can be combined with other analytical data to produce a more realistic analysis of risk.

In fact, when just such information regarding consumer discovery of a defect is included in the risk analysis, the fraction of the defective population ultimately involved in fatality will not approach unity as the number of exposures to the defective product is increased indefinitely, but will approach a smaller fraction instead. McAllister goes on to conclude that when the probability per exposure of product set aside (usage discontinued) due to either discovery of the defect or the experience of a nonfatal injury is high compared to the probability of fatal injury, the probability of fatality per product lifetime may be very small. Undoubtedly, this type of analytical data can benefit the design review and risk evaluation of many new products, especially ones that will be used in critical applications or by their very nature can be potentially harmful to the user.

Utility vs. Risk

In examining the legal aspects of design, Weinstein et al.[8] have concluded that the legal distinction between the reasonably safe product and the unreasonably dangerous one is fundamentally achieved by balancing the product's utility against the potential risks of harm derived from its use. They caution that the courts concern themselves with such factors as:

1. The probability and the seriousness of harm.
2. Burden of precaution against such harm.

3. Foreseeable use (and likely misuse).
4. Causation.
5. Duty.
6. Warnings and standards.

Therefore, the design process should consider these same factors if product liability potential is to be reduced. In other words, we must focus more design attention on the interaction of human behavior and product function because this is a basis for deliberation by the courts in a product liability suit.

There is general consensus that the decision processes involved in designing a new product should consider both sides of the picture: utility and product benefits to the user versus risks the user may encounter. Such a process is illustrated in Fig. 4-3, where we have theoretically balanced the utility and benefits of a product against the potential risks that this same product may pose to the user.

In order to analyze the hazards that a new product design may present to the potential user, Weinstein et al.[9] have revised seven categories originally provided by Wade.[10] These reworked criteria are as follows:

1. Delineate the scope of product uses.
2. Identify the environments within which the product will be used.
3. Describe the user population.
4. Postulate all possible hazards including estimates of probability of occurrence and seriousness of resulting harm.
5. Delineate alternative design features or production techniques including warnings and instructions that can be expected to effectively reduce or eliminate the hazards.
6. Evaluate such alternatives relative to the expected performance standards of the product including the following:
   (a) other hazards that may be introduced by the alternatives.
   (b) their effect on the subsequent usefulness of the product.
   (c) their effect on the ultimate cost of the product.
   (d) a comparison to similar products.
7. Decide which features to include in the final design.

THE ENGINEER AS A MEMBER OF THE DESIGN TEAM

Frequently, in large corporations, the engineer participates in the design of a new product as a member of the design team. A typical design phase for a sizable company is shown schematically in Fig.

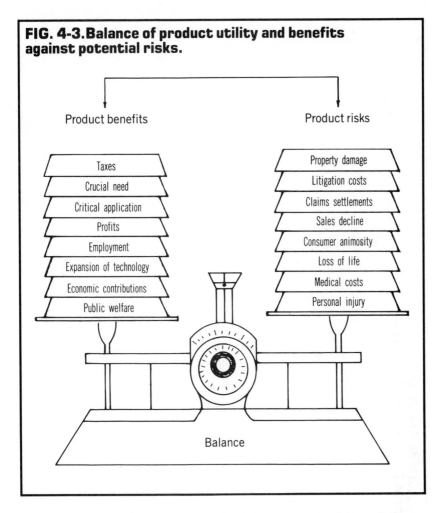

FIG. 4-3.Balance of product utility and benefits against potential risks.

Product benefits

Product risks

| Taxes |
| Crucial need |
| Critical application |
| Profits |
| Employment |
| Expansion of technology |
| Economic contributions |
| Public welfare |

| Property damage |
| Litigation costs |
| Claims settlements |
| Sales decline |
| Consumer animosity |
| Loss of life |
| Medical costs |
| Personal injury |

Balance

4-4. In this role the engineer will integrate his general knowledge or area of expertise with that of personnel from other departments in a company, including other engineering areas, all levels of manufacturing, sales and marketing, purchasing, quality control, planning and production control, and management. For this reason, familiarity with the overall operation of the company and its individual departments is important. Moreover, in addition to completing his specific tasks and assignments, the engineer must cooperate with the team to achieve the best possible overall design and ultimately a functional, profitable product. Yet this type of design approach often overlooks the aspects of product liability introduced by a newly designed product. For example, how many corporate attorneys or legal representa-

FIG. 4-4. Schematic of typical design phase in a large company. (Courtesy of The Travelers Insurance Companies)

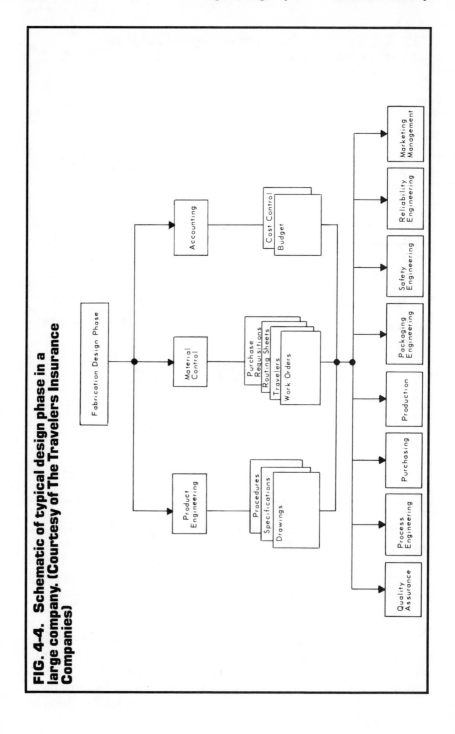

tives are assigned as members of the design team? Or how many are even consulted during the initial planning stages of a new product? Notice that participation of the legal branch is not explicitly listed in the design review chart illustrated in Fig. 4-1. Often, if legal advice is sought, it is an afterthought. Company lawyers (if they do indeed exist) are typically invited to "sit in" on design reviews or analyses of a new product at a later stage, just prior to manufacturing or marketing. It is very likely that their participation in the design of a new product can be beneficial from the viewpoint of product liability; the earlier they are involved and informed, the better.

Nevertheless, the engineer as a resourceful, highly trained technical member of the design team has not only the opportunity but the responsibility to contemplate the aspects of product safety and product liability inherent to the design and manufacture of a product. Competent legal advice may just not be available. Under these circumstances, an engineer may provide the company with valuable insight regarding their product liability exposure. Although we are not advocating or even suggesting the use of an engineer in the capacity of an attorney, we do feel that perhaps he may be in a better position than other members of the design team to recognize potential defects or hazards associated with a design and the concomitant increase in product liability risk such hazards impose. While he and other team members are restricted in the number of alternatives they can propose because of the need to weigh costs, at the same time the engineer must consider what their decisions mean with regard to product safety and the ultimate cost to the company. A favorite expression of the authors which may have applicability to this theme is "First cost is not necessarily final cost." It is penny-wise and pound-foolish to save a few cents or even dollars at the design stage if a manufacturer eventually experiences product liability litigation that runs into the millions of dollars.

Based on some of the examples in Chapter 1 regarding the costs of product liability suits, the above statements concerning the ultimate costs a manufacturer may experience are not the least bit exaggerated. The reader should recall the multimillion-dollar lawsuits involving the Ford Motor Company's Pinto model automobile.

In June 1978, Ford announced that it was recalling 1.5 million 1971-1976 Pintos and 30,000 Mercury Bobcats to improve the safety of their fuel tanks.[11] It is estimated that this recall could cost $40 million or more. In addition, the company also experienced a decline in 1978 Pinto sales, even though these models contain a redesigned fuel tank system, introduced in 1977, which according to Ford eliminates the fire hazard associated with the tank. As a matter of fact, in a recent series of crash tests conducted by the Department of Transportation, and the National Highway Traffic Safety Administration, the Ford Pinto passed the fuel leak test in both front and rear impact accidents.[12,13] These tests specifically determined how well the entire fuel system held up during a crash, avoiding rupture of the fuel tank and/or leakage of fuel.

DESIGN ANALYSIS

The importance of mathematical analyses to the design of virtually any product cannot be overemphasized. These techniques, when accurately conducted, can result in improved product safety and reliability, thus decreasing the product liability potential of a manufacturer. Such analyses must be conducted before sizing the part and selecting the material. Several kinds of mathematical analyses can be administered during the design process, depending upon the particular application or function. These include kinematic, kinetic, stress, weight, and reliability analyses. A brief review of these methods is in order, because each can play a role in producing a safe design.

Kinematic Analysis

Kinematics is the branch of mechanics concerned with the motion of a body, without consideration of the forces required to produce that motion. This subject involves the relationships between displacement, velocity, and acceleration. Kinematics plays a fundamental role in the design of components and systems that involve motion. Needless to say, a huge number of products in today's market contain moving components. As our society progresses, there is a tendency to produce more complex systems. Therefore, the need for this type of design analysis continues to increase.

Kinetic Analysis

Kinetics deals with the action of unbalanced forces on a body and the motion produced by these forces. Kinetics and kinematics together constitute the general area of mechanics known as dynamics. Within this particular branch there are a number of techniques, based on Newtonian mechanics, available for analyzing components subjected to unbalanced forces. The specific method most useful for solving a problem depends on the applied forces and the type of results desired. For example, kinetic analysis by force, mass, and acceleration, by work and energy, and by impulse and momentum are all commonly used in the design analysis of rotary parts and systems.

Stress Analysis

Stress analysis involves the determination of the stresses in a member subjected to mechanical, thermal, and inertial loading. The procedures utilized in stress analysis can vary from the fundamentals in strength of materials to computer-aided finite-element analysis. The topic of stress analysis is discussed in detail in Chapter 5.

Weight Analysis

Weight analysis is becoming increasingly important in this age of fuel and energy concerns. Total weight, the distribution of weight, and material efficiency (high strength-to-weight ratio) should all be considered in the overall design. However, features which promote mobility and energy conservation such as lighter designs and the use of lightweight materials must be considered in view of their possible effects on product safety. Examples of such effects will be found in the section of Chapter 5 dealing with stress analysis, safety factors, and buckling.

Reliability Analysis

With regard to product design, reliability may be defined as the probability (for a certain confidence level) that a component or system will perform its intended function without failure for a given time period under known loading conditions. Although reliability depends heavily on the manufacturing processes, it must also be designed into the product. No wand waving or wishful thinking will instill this characteristic in a product once it's rolling off the assembly line. Reliability, like quality, starts with the initial design and must be considered at all subsequent stages of production. For the interested reader, a detailed treatment of the statistical methods in reliability analysis has been presented by Lloyd and Lipow.[14]

In general, the more complex the design of a component or a system, the greater the likelihood that it will fail prematurely. This is especially true if a system contains many parts or subsystems upon which the operation of the whole product depends. Therefore, the reliability of a component or system can be increased by making its design as simple as performance requirements will permit. A case in point is the Maytag Company of Newton, Iowa, which has been successfully producing home appliances such as washers and dryers under exactly such a concept for years. Their reliability has been confirmed by consistently high ratings in *Consumer Reports*. Presumably, the company has elected to utilize the reliability concept as a marketing tool, a technique which may be used to offset increased manufacturing costs.

The reliability of a new design may also be increased by incorporating parts and materials which have previously demonstrated reliability in service. Designs which depend heavily on the "limits of technology" should be approached with caution. If it is necessary to employ new concepts or ideas, the product should be rigorously analyzed and subjected to accelerated laboratory-field testing as part of development, in order to minimize the risks of product liability problems.

Redundancy

Relative to design, the term redundancy means the incorporation of more than one support path in a structure or component system. Thus, a back-up member exists to carry the load in the event of another member's failure. Scheffey[15] has stated that, typically, designers do not allow for redundant back-up elements in the event that a brittle fracture occurs in a main structural member. Though he feels that it is not economically feasible to insist upon redundant designs, he modifies his viewpoint by adding that where redundancy can be achieved at little additional cost, it should receive consideration.

We would comment on this subject from the product liability aspect. Whether to add the costs required for a redundant design is controversial. The engineering design personnel of a company may feel that a certain redundant design is worthy, while the accounting department cannot economically justify the concept. Frequently economics prevail and a component system or structure is developed which should contain redundant elements, but does not! The question to be addressed is: should the financial analysis be limited to one of manufacturing costs versus profits? Our opinion is that the overall economic picture must consider the costs of both product liability protection and potential litigation in the event of a failure. As we have repeatedly pointed out, the financial satisfactions and related costs in product liability cases often far outweigh the additional costs of initially modifying or changing a design.

Of course, the nature of a particular product or design will heavily influence the details of this consideration, but we caution designers and engineers to seriously contemplate redundancy, not only from a safety and reliability standpoint but also from the aspect of *personal liability.*

REFERENCES

1. Nibel, B. W., and Draper, A. B., *Product Design and Process Engineering*, McGraw-Hill, New York, 1974, p. 5.
2. "Product Quality, Performance and Cost," National Academy of Engineering, Washington, D.C., 1972, p. 5.
3. Lynn, W. R., *Science*, 195, 151 (1977).
4. Furgeson, E. S., *Science*, 197, 834 (1977).
5. "Products Liability and Reliability — Some Management Considerations," Machinery and Allied Products Institute, Washington, D.C., 1967, p. 29.
6. Thornton, P. A., and Colangelo, V. J., "Environmentally Assisted Failures in Ordnance Components," in *Risk and Failure Analysis for Improved Performance and Reliability*, Plenum Press, New York, 1980.
7. McAllister, J. F., "Analysis of Risk Associated with a User-Discoverable Defect," Proc. PLP-77E, New Jersey Institute of Technology, Newark, N.J., 1977, p. 115.
8. Weinstein, A. S., Twerski, A. D., Piehler, H. R., and Donaher, W. A., *Products Liability and the Reasonably Safe Product*, John Wiley and Sons, New York, 1978, p. 139.
9. Ibid., p. 140.
10. Wade, D., "Strict Liability of Manufacturers," 19 *Southwestern Law Journal* 5, 1965.
11. *Newsweek*, June 26, 1978, p. 56.
12. Kramer, L., *Washington Post*, Oct. 15, 1979.
13. *Consumer Reports*, Apr. 1980, p. 222.
14. Lloyd, D. K., and Lipow, M., *Reliability: Management, Methods and Mathematics*, Prentice-Hall, Inc., Englewood Cliffs, N.J., 1962.
15. Scheffey, C. F., "An Analysis of the Point Pleasant Bridge Investigation," in *Structural Failures: Modes, Causes, Responsibilities*, American Society of Civil Engineers, New York, 1973, p. 79.

5

Design Factors Associated With Product Liability

The engineering factors of design which directly affect the product liability position of a manufacturer can be classified into the following main categories:

1. Mechanical and Geometrical Considerations
2. Materials Selection and Specification
3. Utility and Function
4. Unforeseen Service Conditions

Since design-related failures of a very broad range of products leading to product liability litigation can generally be associated with the above categories, it is important to discuss each one thoroughly. Although these four categories necessarily comprise many factors, we will concentrate on those which have critical application to product liability. An outline of the respective factors is given in Fig. 5-1. We hope through this discussion to make it easier for the personnel, technical or otherwise, responsible for these aspects of design and the attendant production to recognize some of the potential dangers, problems, and pitfalls of design-related product liability. We also hope that the subtle relationships between technical elements and product liability problems may become more evident to engineers and the technical community in general.

FIG. 5-1. Outline of design factors which affect product liability.

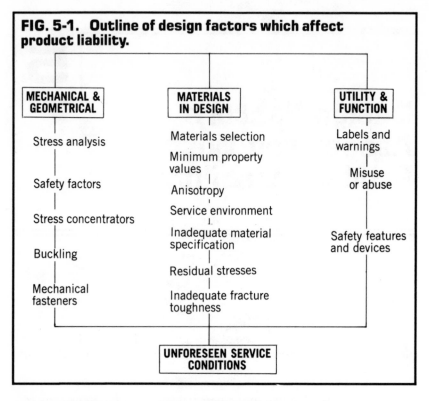

MECHANICAL & GEOMETRICAL	MATERIALS IN DESIGN	UTILITY & FUNCTION
Stress analysis	Materials selection	Labels and warnings
Safety factors	Minimum property values	Misuse or abuse
Stress concentrators	Anisotropy	
Buckling	Service environment	Safety features and devices
Mechanical fasteners	Inadequate material specification	
	Residual stresses	
	Inadequate fracture toughness	

UNFORESEEN SERVICE CONDITIONS

MECHANICAL AND GEOMETRICAL CONSIDERATIONS

Stress Analysis

The design of any component or system is completely acceptable when the designer can reliably predict its performance and establish beforehand the circumstances under which failures will occur. Whether a designer can achieve this somewhat utopian condition depends on many factors, one very important element being a thorough stress analysis. By stress analysis we simply mean a careful determination of the mechanical strength of a component or a structure and its response to the loads applied to it. In situations where extensive field testing is not practical or even possible, accurate stress analysis is an extremely valuable tool in predicting the mechanical behavior of a certain design. Although this approach may not be the best substitute for experience or service, sometimes it is necessary. Radically new designs are examples of such situations. In the case of a new design or untested product, the prediction of failure depends on how accurately the stress analysis reflects actual service or operating conditions.

There are two general problem areas associated with stress analysis that can influence the product liability potential of a newly designed product: (1) an insufficient stress analysis, and (2) an inaccurate stress analysis.

Insufficient Stress Analysis. In most cases, mechanical stress analysis is reliable for those circumstances which are foreseen. In other words, the obvious is relatively easy to deal with. However, the majority of premature design-related failures are due to circumstances which the designer neglects to take into account, either through lack of thoroughness or because of the difficulty of anticipating the loads or forces that may be imposed on the part. Obviously, experience and precision play a strong role in the conduct of such a study.

The interested reader is referred to the summary by the American Society of Civil Engineers[1] of a number of spectacular and catastrophic structural collapses. These studies discuss the details of the particular incidents from a design standpoint, including ponding, bearing, progressive failure, etc., as well as unforeseen loading conditions.

The problem of unforeseen loading conditions was dramatically demonstrated by the epidemic of roof collapses during the unusually heavy winter in the Northeast in 1978.[2] Most noteworthy were the Hartford Coliseum and the C. W. Post Center auditorium of Long Island University. Both roofs were "space frame" designs, consisting of many small steel sections assembled to form a pyramidal network. The damage to the Hartford Coliseum, shown in Fig. 5-2, clearly demonstrates that the roof, not the walls, collapsed. The coincidence between the two collapses raised questions regarding the use of this type of design in new projects. It is generally conceded that space frames, which use an economic minimum of structural material, require more intricate design analysis and are more difficult to construct than the heavier, conventional steel trusses.

Our discussion of space frames is not intended as criticism of this design concept or of the particular stress analyses that were conducted on them prior to construction. Nevertheless, it is an illustrative, widely publicized example of a new design conception

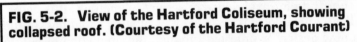

FIG. 5-2. View of the Hartford Coliseum, showing collapsed roof. (Courtesy of the Hartford Courant)

experiencing premature, catastrophic failure. Since the construction of a space frame is considerably more complex and intricate than that of conventional roof trusses, it requires a more comprehensive stress analysis. This may very well be the key phrase: *comprehensive stress analysis.* Generally speaking, the less comprehensive an analytical study is, the greater the risk of product failure and its accompanying product liability consequences.

Could a failure due to insufficient or inadequate stress analysis be interpreted as negligence on the part of the designer or stress analyst and ultimately the manufacturer? The details of a particular case would likely govern such decisions, but the warning should be clear: the stress analysis must be comprehensive. Recall from Chapter 4 the criteria proposed by the legal experts which dealt with the hazards a newly designed product may present to consumers. Several of these categories reflect directly on a poorly conducted or inadequate stress analysis. No doubt, these same criteria would be used as a basis for establishing during a product liability lawsuit whether reasonable care was exercised in the stress analysis. The old advice "To be forewarned is to be forearmed" certainly applies to the engineering personnel involved in these aspects of design and production.

A final thought on this matter concerns the use of state-of-the-art techniques in designing new products: the more innovative the design and the more critical the application (high risk to user), the more exhaustive a stress analysis must be.

Inaccurate Stress Analysis. On the other hand, the stress analysis for any component or system can only be as precise in predicting mechanical behavior as the accuracy of the information and data upon which it is based. Many handbooks are available as aids for computing stress and strain,[3-7] but the most sophisticated analytical techniques in the field of mechanics cannot transform inaccurate and unreliable data into an exact determination of the mechanical response and strength of a design or product. We cannot overemphasize the importance of quality and precision in data gathering for stress analyses.

Recent technological and analytical advances have provided assistance to the stress analyst in conducting both more comprehensive and more exacting studies of the response of a component or system to probable service conditions. These techniques include computer-based analyses such as *finite-element stress analysis,*[8-10] widely available to computer users under the program name of NASTRAN,* and *interactive computer graphics.*[11] Examples of these analytical, computer-based tools are shown in Fig. 5-3 for a ribbed baseplate component. Not only have these methods and related techniques made the stress analyst's job easier, but their tremendous value to the overall field of engineering design and analysis is just beginning to be appreciated as they expand the capabilities of the analyst and enable more accurate stress analysis.

Along with structural engineers, engineers in the automotive and aircraft industries have been quick to recognize the value of applying finite-element and interactive graphics to stress analysis as well as inherent design. Through these techniques, the designer or engineer can immediately see (on a video display) the response or changes in a design or system as new parameters (e.g., load changes) are introduced in a program. We strongly suggest an interactive graphics demonstration for any engineer or designer not familiar with this tool.

---

*The NASTRAN finite-element stress analysis program is available from "COSMIC," Barrow Hall, University of Georgia, Athens, Ga.

FIG. 5-3. Example of interactive computer graphics for a baseplate. (Courtesy of J. Pascale)

(a) Overall view of baseplate

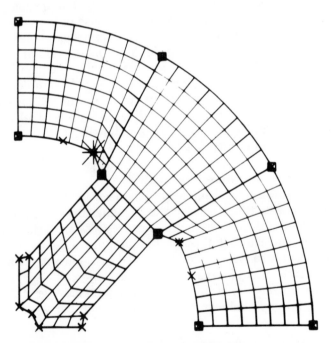

(b) Computer graphics plot of finite-element grid representing one quadrant

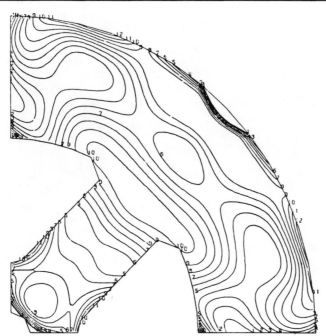

(c) Stress contour plot of same quadrant as in **(b)**

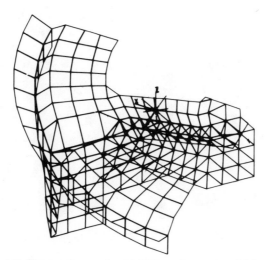

(d) Three-dimensional finite-element grid for the subject quadrant

FIG. 5-4. Failed automotive fan blade.

(a) Overall view of blade

(b) Cracking in the failed blade (arrow)

The importance of stress analyses in design can perhaps best be demonstrated by example. An automotive fan blade failed unexpectedly, causing serious injuries to a mechanic who was working on the engine. The water pump shaft and its bearing assembly also failed in conjunction with the fan blade. The fan blade is shown in Fig. 5-4(a). Subsequent investigation revealed that fatigue cracks had initiated in the fan blade at the region where the two blades were joined, as illustrated in Fig. 5-4(b). It should also be noted that this was the second incident involving the same fan blade design and manufacturer, in a relatively short period of time.

A materials examination of the blade and water pump shaft disclosed no defects or material deficiencies. Based on the available evidence, the cracking and failure of the fan very likely produced an eccentric loading condition on the water pump which probably led to its failure.

Subsequent to the production of the original blade, the fan blade manufacturer added a stiffener to the blade assembly as shown in Fig. 5-5. This addition created a "sandwich" effect, thus distributing the loads over a larger area. The net result would be a decrease in operating stresses at the subject location, thereby reducing the tendency for fatigue failures. Does this incident constitute a design

FIG. 5-5. Modified fan blade with stiffener plate added (arrow).

failure? This point was never argued in court. The attendant product liability litigation included the manufacturers of the fan blade, the water pump, and the water pump bearings. Although the failure of the fan blade probably precipitated the failures of the water pump, all three manufacturers shared in the costs of an out-of-court settlement to the plaintiff.

The conditions of loading and design which caused this kind of failure in the subject fan blade might have been prevented by the application of interactive graphics or a finite-element analysis during the design stages of production.

Finally, a complete stress analysis should include a study of the effect not only of static and dynamic applied loads from mechanical and thermal forces but also of the internal stresses resulting from forming and fabrication processes (residual stress). The topic of internal residual stress is treated later in this section. Furthermore, residual stress is such an important aspect of product performance and reliability that it will continually be mentioned and discussed throughout Chapter 7.

Safety Factors

Although this will come as no surprise to the practicing metallurgist and materials engineer, all engineering materials exhibit variability in mechanical properties. This variation is influenced by virtually all stages of manufacturing, as we will see in Chapter 7. From a design standpoint, the inherent problem with variability is compounded because uncertainties can exist regarding the magnitude and direction of applied loads. This means that approximations are sometimes necessary in calculating applied stresses for all but the most simple members. Therefore a prudent designer allows a contingency for the possibility of accidental or abnormal loads of unusually high magnitude. Could such loading have contributed to the roof collapses cited earlier in this chapter? It certainly must be considered a possibility.

Accordingly, to insure a margin of safety and thus protect against failure from unpredictable stresses, it is necessary that the allowable stresses in a member or system be smaller than the stresses which cause failure. Dieter[12] has defined the working stress (allowable stress) to be the value of stress for a specific material used in a particular manner which is considered to be a safe stress. This working stress (σ_w) is usually based on the yield strength (σ_{YS}) in the case of ductile metals and the ultimate tensile strength (σ_{UTS}) for brittle metals. The values of working stress are thus expressed as follows:

$$\sigma_W = \frac{\sigma_{YS}}{N} \qquad \text{Ductile metals}$$

$$\sigma_W = \frac{\sigma_{UTS}}{N} \qquad \text{Brittle metals}$$

In these expressions, N is the safety factor. Thus, it is apparent that N is the ratio of the strength of the material (yield or ultimate) to the allowable stress. This relationship is shown below for a ductile metal:

$$N = \frac{\sigma_{YS}}{\sigma_W}$$

Hence, the problem is one of determining the proper value of N. If N is unnecessarily large, we are probably using more material or higher strength material than is needed for the application. In other words, the component or member is overdesigned. This may be the conservative approach from a safety position, but it also may be economically unfeasible.

Values of working stresses are established by local and federal agencies and by technical organizations or societies such as the American Society of Mechanical Engineers (ASME). Working stresses may be specified in codes for certain applications, such as ASME Boiler Construction Code, Code for Pressure Piping, Code for Power Shafting, etc.

Overall, the value of N depends on consideration of all the factors previously mentioned, plus careful consideration of the consequences of a premature failure. Factors of safety based on yield strength are often taken between 1.5 and 4.0.[13] For more reliable materials or familiar design and service conditions, the lower end of this range is appropriate. For untried materials and design or uncertain loading conditions the higher factors are safer. In any event, if a failure could result in loss of life an increased safety factor should be considered.

. Arbitrarily increasing the factor of safety in connection with a questionable design or component is not sound engineering. If the available data and background information are inadequate to determine a dependable safety factor, then perhaps a restatement or reformulation of the original problem is warranted.

Stress Concentrators

Geometric changes in stress-carrying components and members can produce nonuniform stress distributions and locally high stresses in the vicinity of the change or discontinuity. Such discontinuities may take the form of sharp radii in fillets, sharp corners, changes in cross section, keyways, chamfers, bosses, holes, etc. These geometrical alterations can exist both internally or externally (on the surface of a part) depending on the specific design. Unfortunately, just about any variation in geometrical continuity will result in stress concentration (stress raiser). The more abrupt the change, the greater its stress-concentrating effect. Stress concentration is generally expressed as the ratio of the maximum stress to the nominal stress, based on the net cross section. This nominal stress is often referred to as the engineering stress (load divided by area). The stress concentration factor (K) is then expressed as follows:

$$K = \frac{\sigma_{max}}{\sigma_{nominal}}$$

Some of the more familiar and easily calculated stress concentrators are shown in Fig. 5-6.[14]

The effect of a stress concentrator in a component or member of a system, therefore, is to promote premature, usually unanticipated, failure. Stress concentrators themselves may be completely responsible for a failure or they may just provide the initiating condition. The eventual failure mechanism may then take any number of forms, which will be discussed in Chapter 6. The important point is that abrupt geometric changes in design produce stress raisers, which can seriously affect the performance of a product and thus influence its product liability potential. As we have frequently emphasized, the design stage is a most appropriate time to take such factors into consideration in order to minimize the risks associated with them and avoid future product liability problems.

As an example, we have illustrated a common stress concentration condition in Fig. 5-7. One version of the design contains a very sharp radius on the inside corner. Notice how the lines of force representing the stress distribution concentrate in the vicinity of this abrupt geometrical change. A designer can easily alleviate this condition early in the design stage by specifying as generous or large a radius as practical. The corresponding stress distribution for such an alternative is also illustrated in the figure. The deleterious effect of the stress raiser has been all but eliminated.

A convenient method for displaying stressed areas and stress distributions in components is to transmit polarized light through plastic (polymer) models of the parts in simulated service. Stressed regions will not transmit as much light and therefore will be revealed by contrast. Brittle coating stress analysis techniques can also be useful for providing a graphic picture of the distribution, direction, location, and magnitude of tensile strains, particularly in geometrically complex parts.

The force-flow concept is frequently used to visualize stress concentration and the distribution of stress in a part subjected to applied forces.[15] Although the stress concentration condition may also depend on how the component was formed or fabricated, the lines of force tend to concentrate at locations with sudden geometric changes. If at all possible, it is good practice to locate these discontinuities in regions of the component subjected to lower stresses. However, such a design modification may not be feasible or even possible. In this case, perhaps it may be feasible to incorporate other geometric alterations in the design, such as holes and grooves, to improve the stress distribution in the vicinity of the stress raiser. Following our discussion of the deleterious effects that geometrical changes have on the stress distribution in a component, such a suggestion may seem absurd or at the very least contradictory. But consider the alternatives presented by Chow[16] in Fig. 5-8. He shows that the introduction of holes and grooves can alleviate stress concentration effects of some discontinuities and that in some instances different materials can be used to modify the stress transfer between two components.

Buckling

The designer of a structure and certain engineering components must necessarily consider the possibility of buckling. This type of failure, sometimes referred to as an instability failure, can occur under several loading conditions, including compression, bending, and torsion. In 1757 Euler described buckling in terms of a critical value of the compressive force which produces large lateral deflections in a slender column or strut on application of the slightest lateral load. This critical load for a pin-connected column is expressed as follows:

$$P_{cr} = \frac{\Pi^2 EI}{l^2}$$

where E is the modulus of elasticity; I is the smaller principal moment of inertia of the cross section; and l is the length of the bar or column.

When the value of compressive force reaches this critical load, the lateral deflection is usually sufficient to cause complete failure of

FIG. 5-6. Theoretical stress concentration factors for various geometrical discontinuities. (Ref 14)

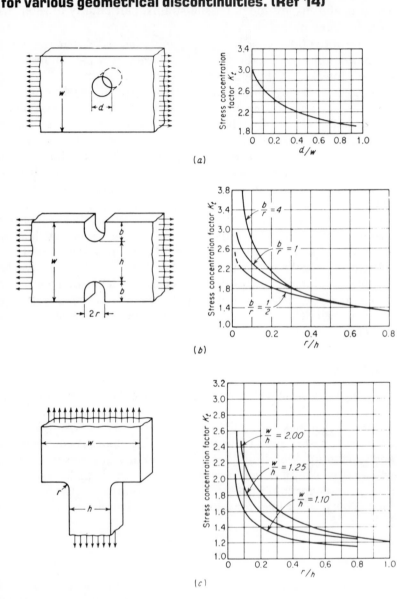

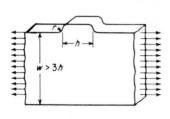

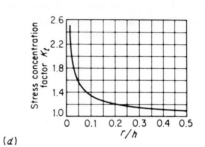

(d)

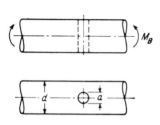

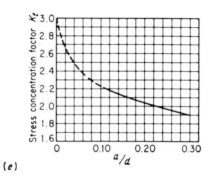

(e)

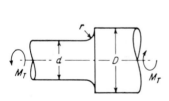

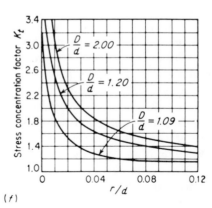

(f)

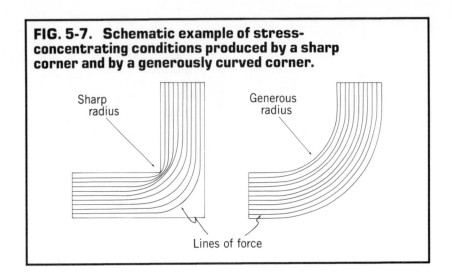

FIG. 5-7. Schematic example of stress-concentrating conditions produced by a sharp corner and by a generously curved corner.

Sharp radius

Generous radius

Lines of force

the member or structure. The important point here is that the critical value of load does not depend upon the strength of the material but rather on the structural dimensions and the modulus of elasticity of the material (see the above equation). Timoshenko further points out that two equal-sized slender struts, one made of a high strength steel, the other of a common structural steel, will buckle at the same compressive load even though the strength of the two steels differs significantly.[17]

Thus, to resist failure by buckling, the members of a structure or a system must not only be strong enough to carry the applied loads but must also be rigid enough to remain stable. The stability of long, slender members is very dependent on the modulus of elasticity, a measure of the stiffness of the material. The modulus of elasticity for several engineering materials is given in Table 5-1.[18]

Several examples of this instability mode of failure are illustrated in Fig. 5-9 through 5-11. Buckling in narrow beams is particularly interesting from the standpoint of product liability. Consider the design of a simple symmetric beam; from basic mechanics we know that the maximum bending stress (flexure stress) in the outer fibers of a beam can be changed by varying its geometric shape (or by changing its orientation with respect to the applied load). Such a situation is illustrated in Fig. 5-12, for simple bending. Examining this figure, we see that for the same cross-sectional area of beam (therefore the same amount of material), the beam in case 2 experiences a maximum bending stress four times greater than its counterpart in case 1. There-

TABLE 5-1. Modulus of Elasticity for Several Engineering Materials

Material	Modulus of elasticity (E) 10^4 MPa	10^6 psi
Gray cast iron...........	10	15
Steel....................	20	30
Aluminum	6.9	10
Brass (70% Cu)	9.7	14
Wood (Ref 19)	0.69-1.4	1-2

FIG. 5-8. Geometric techniques for alleviating stress concentration effects of certain discontinuities. (Ref 16)

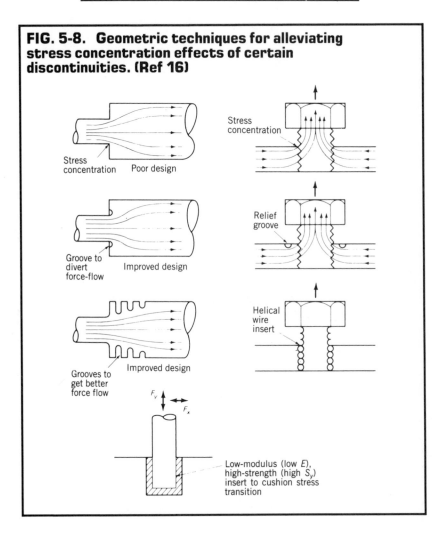

FIG. 5-9. Schematic exam- **FIG. 5-10. Buckling of a**
ple of buckling in a column **cylindrical shaft subject-**
subjected to compressive **ed to torsion. (M = moment)**
loading.

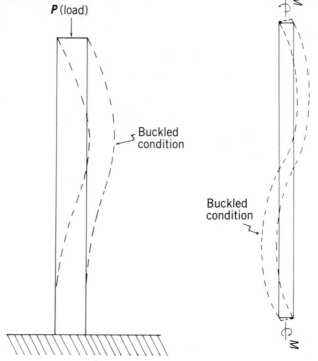

FIG. 11. Examples of buckling failures in several
structural shapes and loading conditions. (Ref 16)

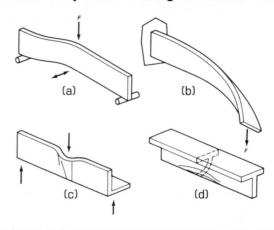

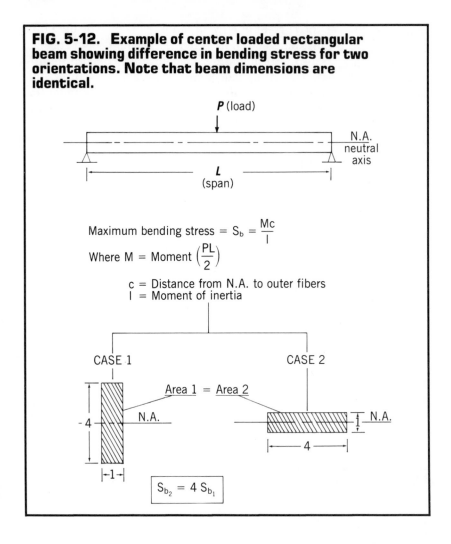

FIG. 5-12. Example of center loaded rectangular beam showing difference in bending stress for two orientations. Note that beam dimensions are identical.

P (load)

N.A.
neutral
axis

L
(span)

Maximum bending stress $= S_b = \dfrac{Mc}{I}$

Where $M = $ Moment $\left(\dfrac{PL}{2}\right)$

$c = $ Distance from N.A. to outer fibers
$I = $ Moment of inertia

CASE 1

Area 1 = Area 2

CASE 2

N.A.

N.A.

$S_{b_2} = 4\,S_{b_1}$

fore, if we can increase the efficiency of a beam (its load-carrying capacity) just by rotating it 90°, in essence making it "deeper," why not design all beams this way? Within certain limits such a technique is acceptable and very useful. However, as Chow[20] appropriately points out, a beam that is very deep and consequently very thin may not easily fail by yielding (maximum bending stress exceeds the yield strength), but would buckle rather easily. As stated previously, buckling results from instability due to insufficient stiffness, and thin, slender members are prone to this condition. Unfortunately, we tend to design structures with thin sections in order to utilize materials more efficiently and therefore more economically. In the case of high strength-to-weight applications, a concept becoming more and more attractive as the current energy situation worsens, the buckling mode of failure must be given more attention in the design stage.

Since the possibility of buckling is a cause for concern in efficiently or economically designed components and structures, the designer who wishes to avoid this type of failure and its potential product liability risks should be familiar with the techniques for preventing buckling. A number of these methods are summarized as follows:[21]

1. Use a more rigid material, higher modulus of elasticity (E), and higher shear modulus (G).
2. If feasible, employ a lower yield strength material so that yielding rather than buckling predominates. High strength materials will eventually fail by buckling because yielding does not occur.
3. In tubing applications, increase the wall thickness. A small-diameter tube with a thick wall of ductile material is less likely to buckle under bending conditions.
4. Incorporate lateral support to prevent deflection in the bending mode.
5. Utilize stiffeners such as ribs, corrugations, honeycomb structures, and surface textures to prevent localized buckling. The use of ribs to prevent buckling in the web portion of angle beams is shown in Fig. 5-13.
6. Use a filler material such as sand, wax, resin, or even water when bending thin-wall tubing. The filler stiffens the walls by hydrostatic pressure during the bending operation. After bending is complete, the filler material can be reclaimed. This concept

FIG. 5-13. Example of a stiffened member which resists buckling. (Ref 16)

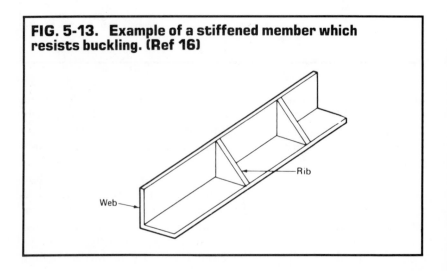

Rib

Web

has application in structures in which one wishes to prevent ovality during the bending of thin-wall tubing.

Captive Columns

A unique technique for avoiding buckling while increasing the strength-to-weight ratio of columns is the "captive column" concept.[22] In the column designed by Bosch (U.S. Patent No. 3,501,880), shown in Fig. 5-14, the three individual components are stressed in their most favorable (strongest) directions as follows:

1. The cores are placed in radial compression.
2. The columns are placed in axial compression.
3. The filaments are placed in tension.

This particular combination of structural members restricts buckling while exhibiting a very high strength-to-weight ratio. According to

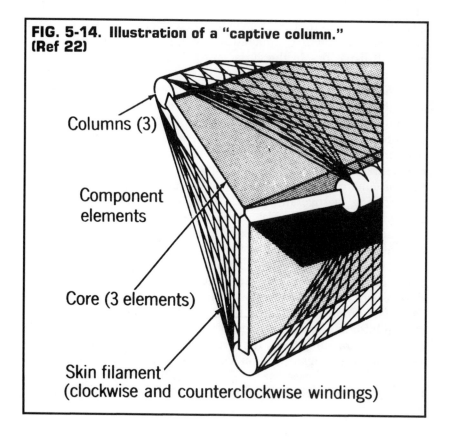

FIG. 5-14. Illustration of a "captive column." (Ref 22)

Columns (3)

Component elements

Core (3 elements)

Skin filament (clockwise and counterclockwise windings)

the inventor, the main load-carrying column elements are supported along their entire length so they cannot buckle or move in any direction. The individual columns are held captive by wrapping them tightly against a central core with a high strength, uniaxial material such as fiberglass.This could be accomplished relatively easily by means of filament winding. The net result is that the three basic elements of the captive column work together to resist any type of loading applied to the structure.

Balanced Designs

Reinforcing certain structural materials is sometimes desirable in order to take advantage of the beneficial characteristics of a particular material. This technique can be very economical from both the weight and the cost standpoint. In fact, many varieties of fiber-reinforced materials (metals and polymers) have resulted from this concept. However, in order for the composite system to function properly, the interfacial bond between its constituents must be strong, providing for adequate stress transfer from matrix to reinforcing fiber. This bond is enhanced by the use of chemical wetting agents and also by mechanical disparities in the surface of the fibers, or corrugations as in the case of steel reinforcing bars in concrete.

The weak tensile strength of concrete is well known, so steel bars are embedded in this structural material to increase its tensile loading capacity. Economy also results, since less concrete is necessary when reinforcing steel rods are used. If a concrete beam is placed in bending, for example, the tensile stress in the portion of the beam below the neutral axis is principally sustained by the steel bar. The proportion of concrete in this section can therefore be reduced without jeopardizing the overall strength of the composite beam. Such an example is illustrated in Fig. 5-15. This technique is characterized as "balancing designs." By balanced design we mean that the member is designed to produce the same stress in the matrix and the reinforcing phase. Correspondingly, a balanced failure is one in which the matrix and reinforcing phase fail simultaneously. Unfortunately, balanced designs can also result in catastrophic failures. For instance, in the case of a reinforced concrete structure, if the concrete phase and the steel reinforcing bar are balanced with respect to ultimate tensile strength or fracture stress, the structure can fail suddenly with no visible warning signs. A safer condition exists if the steel bar is purposely understressed at the point of failure of the concrete. This technique, known as overreinforcement, will permit visible signs of the impending failure before the total collapse of the structure or member. Such techniques are designed to reduce the potential hazards associated with sudden structural failures. As a consequence, they have a beneficial effect on the product liability position of the responsible design personnel and their company.

"Snap-Through" Instability

A final word regarding instability-type failures. A mode similar to buckling can occur when an arched member or component such as a dome or shell is loaded to the point where it deflects to a horizontal position (straightens out). At this critical value of load, the member suddenly snaps through the straight position and takes up the opposite curvature. Because this phenomenon is frequently observed in the ends of drums and closed-end cylinders, it is referred to as "oil-canning."

MECHANICAL FASTENERS

The construction and assembly of most components and structures depend heavily on mechanical joints as well as metallurgical

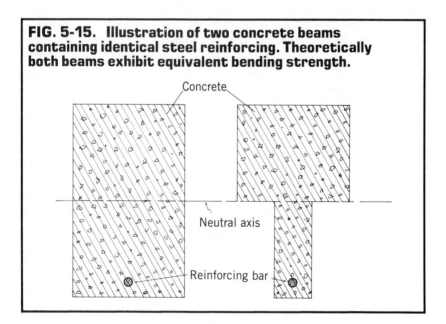

FIG. 5-15. Illustration of two concrete beams containing identical steel reinforcing. Theoretically both beams exhibit equivalent bending strength.

Concrete

Neutral axis

Reinforcing bar

joining. The various techniques and product liability ramifications of metallurgical joining such as welding or brazing will be analyzed later in the discussion of processing in Chapter 8. Here we will concentrate on mechanical fasteners such as nuts and bolts, rivets, adhesive bonding, etc. Obviously, there is a plethora of these devices, from the commonplace carpentry nail to the most complex special purpose fasteners. In fact, an annual issue of *Machine Design* is devoted entirely to just fastening and joining.[23] Since it is not possible or practical to cover all these fasteners, we will attempt to point out the particularly troublesome fasteners, the ones more susceptible to failure. The importance of fasteners may be due to the frequency of their use and also the critical nature of their application.

In the area of permanent fasteners, for example, the common carpentry nail comes in various sizes and shapes and is used literally by the millions. Perhaps because of familiarity this type of fastener is widely taken for granted. However, the ability of a nail to maintain joint integrity depends not only on the strength of the nail but also on the contact area between the nail and the material it is fastening. Thus, the length and diameter are important as well as the number of nails used to "make a joint." In other words, fastening materials with nails, particularly in large, heavy structures, should not be taken lightly. The failure of these fasteners can have both economic and safety consequences. For instance, we are aware of a series of incidents involving the failure of nails in the roof framing of new residential housing. The common ingredient in all the cases was nail pullout in the rafters and ridge, to the point where these joints were seriously weakened. In a relatively short period of time (months), certain joints failed and eventually the roofs sagged under their own weight. A few actually collapsed under light loads of snow. Although in these cases the framing lumber was apparently not dimensionally stable, it was also concluded that there were insufficient nails to make a proper joint! The builder-contractor (designer) spent considerable sums to repair and replace the defective roofs and inspect all roof framing in the tract-type housing development. A similar problem shows up in new flooring where the underlayment tends to pull up at joints and eventually breaks through the floor-covering material. This failure is frequently attributed to an insufficient number of flooring nails.

Domino Effect

Generally the design of a component or a structure is predicated on the factors we have previously discussed — stress analysis, safety factors, etc. Occasionally the integrity of an entire structure may depend heavily on the performance of a single fastener or joint. Although this is poor design practice, sudden, catastrophic failures are a constant reminder that it exists. The reader is reminded of the incident involving collapse of the scaffolding at a cooling tower construction site, cited in Chapter 1. Think for a moment what the outcome might have been if the top scaffolding had been not continuous but segmented, with sequential breaks in the construction. However, to pinpoint the exact cause of such consequential failures on a single fastener is usually extremely difficult; very often the failure of one fastener serves as the initiating event in a sequence of failures that ultimately ends in complete failure of the structure or system. Since this process, also known as progressive failure, is analogous to the toppling of dominos, it is called the "domino effect." For example, the bolts in a certain structure are designed to support a given stress level. If only one bolt fails, the structure is likely to remain intact. But the remaining functioning bolts now share the load of the failed member, the nearest neighbors taking a greater share of this load. Systematically one or more of the remaining bolts may experience a stress in excess of their yield strength. Yielding can eventually result in further failures. This process keeps repeating at an increasing rate until the structure is no longer self-supporting and undergoes an apparent "sudden" failure or collapse.

Rivets

Riveting is a very common method of assembling metal parts and structures which can be adapted to production line operations. A large aircraft, for example, can contain hundreds of thousands of rivets. Rivets are a permanent type of fastener because they are permanently deformed on the protruding end by a process called heading. Heading can be accomplished when the rivet is either hot or cold. In the case of aluminum, the deformation is performed at ambient temperatures because cold working increases the strength and hardness of aluminum. The actual operation may be accomplished by hammering or by the application of pressure.

Although rivets are usually stronger than threaded fasteners of comparable size, if the flange (or head) of the rivet is improperly formed it will tend to fail quickly under applied stresses. As a rule of thumb, the rivet should protrude from the material to be fastened by a minimum of 50-70% of its diameter in order to produce a proper joint when headed. In spite of any recommendations regarding the length of rivets, it is still considered good design practice to determine rivet length by experiment, favoring an overlong rather than too short a rivet.[24] If no rotation is desired, the flange or head should be brought into firm, flat contact with the pieces to be fastened.

Whenever possible, riveted joints should be designed so the rivets are not placed in tension. Otherwise, a slight eccentricity of the applied load exerts a lever action on the head or flange that can result in premature failure. This tendency for early failure is promoted by repeated or cyclic loading.

When the applied loads are oriented normal to the line of rivets and to the rivet axis, these fasteners may fail as follows:

1. The rivets may simply shear off as illustrated in Fig. 5-16(a).
2. The plate may crack or tear at the cross section through the rivet hole, as shown in Fig. 5-16(b).
3. The plate may deform around the rivet, enlarging the hole. This is known as a bearing failure and is illustrated in Fig. 5-16(c).
4. The rivet itself may also be crushed.

Furthermore, in an improperly designed joint, failure may also occur through tearing of the edge of the plate in back of the rivet hole and/or shear failure of the plate behind the rivet. These modes of failure are unlikely to occur if the distance from the center of the rivet to the plate edge is twice the diameter of the rivet.

More specifically, problems with the driven head or flange of rivets can include eccentricity, nonuniform thickness, and improper seating with the pieces to be fastened. Distortion can result in the fastened materials because of poor initial contact (lack of clamping), improper riveting sequence, and excessive heading force. When many rivets are installed in a row, it is better to start at the two ends, then the middle, then fill in between at wide spacings. The gaps are finally filled in sequentially. This procedure distributes any distortion more uniformly along the entire joint.

It is always good design practice to use rivets (as well as any other metallic fasteners) made from the same alloy or material as the pieces to be fastened. This avoids the possibility of galvanic corrosion, as we point out in Chapter 6, and assures that the rivets and pieces to be fastened are comparable in hardness and strength. It is poor practice to drive a hard rivet in a soft plate.[25] For instance, if a high strength aluminum rivet is cold-driven into a low strength aluminum plate, the plate may undergo distortion.

Finally, when rivets are loaded parallel to their axis in compression, they may be subject to buckling failures. The shank portion can buckle because of insufficient strength or excessive clearance in the

FIG. 5-16.   Schematic example of rivet failures.

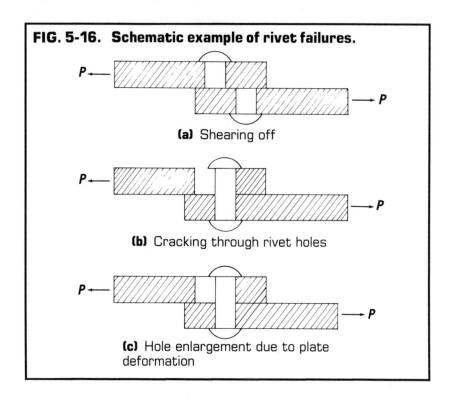

(a) Shearing off

(b) Cracking through rivet holes

(c) Hole enlargement due to plate deformation

plate holes. Tight holes will provide lateral support and prevent buckling in solid rivets. However, in the case of tubular, split, or bifurcated rivets there is no internal support and buckling can occur.

Nuts and Bolts

Nut and bolt combinations are certainly one of the most common and widely used fastening devices in joining materials. They are typically fabricated from a range of steels — carbon and alloy, stainless steels, and alloys of aluminum, copper, and nickel. These fastening devices are a carefully engineered product which tend to be taken for granted and are frequently used without proper consideration for the overall design concept.

For example, an ordinary nut can become loose when the forces of vibration exceed the friction force between the engaged threads. Under such circumstances, the bolt can be subjected to eccentric loads, bending stresses, and repeated loads which can eventually produce a premature failure. This is of course a very undesirable situation, so there are many methods to prevent nuts from working loose. These techniques include:

1. Peening the exposed end of the bolt.
2. Deforming threads of the nut.
3. Using lock washers.
4. Using special threads on both nut and bolt.
5. Sealing the threads with adhesive.
6. Employing various types of locknuts.

Lock washers are used primarily in static situations where a change in length of the tightened system is possible through wear, thermal expansion, or creep. If vibration is the problem, however, lock washers are not particularly effective.

Bolt Strength

As Chow points out, from the standpoint of strength, bolts should be tightened as much as possible.[26] This places the bolt in tension and the nut in compression. Yet, from a practical viewpoint, this recommendation must be approached with some caution. First, anyone who has inadvertently overtightened a bolt or machine screw to the point of failure will attest to the frustration and extra work involved in extracting the stuck portion using an "easy-out" (screw extraction tool). Secondly, many nut/bolt systems are designed to specific torque criteria. In this case, too much or too little tightening is deleterious and may contribute to a premature failure. Among the potential failure modes are fatigue and environmentally assisted cracking such as hydrogen embrittlement and stress corrosion cracking. For example, the nut on the spindle assembly of a 175 mm breech mechanism was inadvertently tightened in excess of its torque specifications.[27] The area of the failure in the threaded spindle and its relationship to the assembly are shown in Fig. 5-17. Overtightening resulted in a tensile stress in the shank which when combined with an environmental attack produced stress corrosion cracking. On one occasion, complete failure and expulsion of the rear portions of the spindle occurred during operation of the system.

In another example, a bolt in a structural steel girder system for an overhead crane became loose and eventually failed in fatigue. The broken bolt is shown in Fig. 5-18. Fortunately the bolt was discovered on the shop floor by a conscientious employee who reported the finding to his supervisor and thus its identity was established. During the ensuing investigation, a number of other loose bolts were discov-

FIG. 5-17. Threaded spindle component.

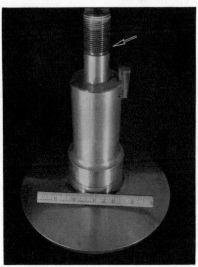

(a) Overall view of component (arrow denotes region of failure); scale in inches

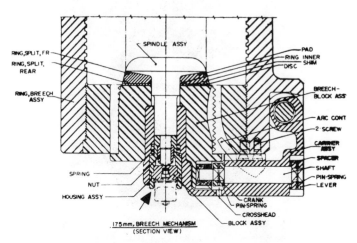

(b) Sectional view of component in assembly; arrow shows portions which were expelled during operation

FIG. 5-18. Failed bolt from steel girder system supporting an overhead crane.

(a) View of failed bolt; note mechanical damage to shank portion

(b) View of fracture surface exhibiting reversed bending fatigue features

ered in the girder assembly and the situation was rectified before a serious product liability situation developed.

Consequently, proper tightening procedures should be established. Volume 10 of the ASM *Metals Handbook*[28] contains an article describing the common modes of failure for bolts and the rationale usually used to establish preloads on the bolts.

The designer is in the best position to conduct a stress analysis and determine the proper degree of tension. He should insure that bolt torque requirements are clearly specified on the original assembly and upon reassembly if the component to be fastened is one that is commonly removed or replaced. Failure to take this precaution can result in a situation of potential liability should a failure of the bolts occur.

Adhesive Bonding

Adhesive bonding has recently become more important and more practical as a method of fastening components and structures. This is due in part to improvements in polymer technology and the demand for high strength-to-weight ratio materials in assembly. Basically there are two types of bonding mechanisms:[29]

1. Adhesion resulting from interatomic or intermolecular action between the adhesive and the assembly.
2. Mechanical adhesion produced by the penetration of the adhesive into the asperities of the assembly surfaces.

Any condition that interferes with these two mechanisms will inherently degrade the bonding action and contribute to failure of the joints. In the case of interatomic or molecular attraction, the integrity of the bond is affected by such factors as pressure, temperature and time during curing, entrapped moisture and air, contamination of the adhesive, aging, etc. For the case of mechanical adhesion, surface preparation is very important. Any contamination of the assembly surfaces, from dirt, grease, oils, water, and even oxides, can seriously reduce the effectiveness of the bond, thus leading to early failure of the joint.

The designer who considers or utilizes adhesives for fastening and joining must be familiar with the unique chemical, physical, and mechanical properties of these materials. For instance, the thermoplastics, including acrylates, styrenes, and vinyl chlorides, soften upon heating. The thermosetting resins such as epoxies, polyesters, and phenolics, on the other hand, cure and become hard with the application of heat. Unfortunately, most adhesives have a limited temperature range in which they can operate safely and reliably.

It has been proposed by Nibel and Draper that poor engineering design is responsible for most of the failures in adhesive joints.[30] If their indictment of design engineers is legitimate, then their recommendations for avoiding failures and thus precluding product liability problems are well worth mentioning:

1. The assembly or structure should be designed to allow a large enough area of adhesive bond to support the applied shear loads.
2. The joint and thus the adhesive bond should be stressed in shear rather than tension.
3. The joints should be tapered for greater strength. Both criteria 2 and 3 suggest that butt joints should be avoided whenever possible, and that lap joints, particularly tapered lap joints, are better design practice.
4. The adhesive should be applied in uniform thin layers to avoid irregularities and stress concentration.
5. Combinations of materials should be selected so they are compatible from the standpoint of thermal expansion.

MATERIALS IN DESIGN

Any good design must incorporate materials selection. This is necessary because the total performance of a product requires the integration of materials, geometrical design, and operating conditions. In order to optimize a design, materials selection should begin early in the process. Thus the desirable attributes of a material can be maximized while its undesirable features are minimized. This view clearly denies the premise that design is concerned with geometric shape only and that materials selection is a secondary consideration and a separate process. In some instances, the selection of a material for a design is, unfortunately, an afterthought. Such an approach to materials selection may eventually result in material failures which are design related.

The working engineer or technical person not familiar with product liability may not grasp the difference between a material failure that is design related and one that is not. In this section we will give various examples of materials failures that are design related, while in Chapter 7 and 8 we will discuss materials defects or failures that are production related.

Materials Selection

As previously implied, a design must ultimately be reduced to a tangible piece of hardware. In this process, the materials of construction are selected which should perform satisfactorily under the anticipated service conditions. Often, however, errors occur in the selection process whereby an improper material is incorporated into the finished product. We will assume for the purpose of this discussion that the designers are reasonably competent and therefore we will not include the situations where there has been a gross misapplication of a material in a design—for example, the use of a thermoplastic resin in an elevated temperature application or the use of a brittle material like cast iron in a shock-loading situation. Neglecting these obvious material-related design defects, what kinds of errors can occur? The following is a tabulation of the more common, not so obvious, sources of defects or errors:

Failure to specify the minimum level of properties which are acceptable (material variation).

Failure to consider anisotropic conditions in a material.

Failure to properly anticipate the service environments, such as wear, corrosion, and vibration, and the operating conditions, such as fatigue.

Inadequate materials specification.

Failure to consider residual stresses.

Utilization of material with inadequate fracture toughness.

Minimum Property Values

Failure to appreciate the statistical nature of materials is an occurrence which usually affects small companies without extensive materials engineering departments rather than companies which employ extensive staffs. However, in some instances, i.e., new materials areas, it can affect even these large organizations. Virtually all alloys when tested, for example, will show a variation in mechanical properties from heat (lot) to heat or even within heats, resulting in a statistical distribution. Such observed variation is due to real variation in the material itself plus variation from experimental error in the tests. Often an inexperienced designer will design around the typical value for an alloy, not realizing that the lowest value which may be obtained from a series of heats is usually considerably lower than the typical (modal) value which is published in the literature. It is the minimum expected value for a particular property which should be used as a basis for design, since this material will most assuredly find its way into service. To act otherwise is to reduce the factor of safety and invite premature failure.

Anisotropy

In many mechanical engineering texts, metals are presented as being isotropic and continuous with respect to their mechanical properties. In reality, metallic alloys ordinarily contain dissolved and combined gases, nonmetallic inclusions, and variations in the local composition referred to as chemical segregation. These defects, especially compositional variations and inclusions, together with the shaping process such as rolling or forging, combine to produce a variability in the mechanical properties, particularly ductility and toughness (and sometimes the corrosion rate), which is dependent upon the direction of mechanical working. This variability called anisotropy must be considered when choosing an alloy for a design. For instance, if the design is predicated on the longitudinal properties of a material and the maximum applied stresses are in the transverse or short transverse direction,[31] that design may be jeopardized.

Such failures have occurred and they have occurred in substantial members. One of the more prominent technical areas where anisotropy has played a role in failure involves some large welded steel structural members. In 1973, two major building projects,[32] a performing arts center in El Paso, Texas, and a power transmission line at Grand Coulee Dam, were delayed with the result that considerable financial losses were incurred. The cause? Lamellar tearing resulting from poor ductility in the "through plate" or short transverse direction. The cracks occur when molten weld metal deposited on the surface of the structural member cools. When the joint is restrained, stresses result. Ordinarily, the material is sufficiently ductile so that it yields locally and relieves these stresses. However, in the examples referred to above, the material ductility was insufficient in certain orientations because of anisotropic conditions in the large plate sections. When the concentrated stress reached a weak plane in the material, cracking initiated. The solution? To educate structural engineers and designers that metals and other structural materials do not usually exhibit the same properties in all directions.

Additional examples of anisotropic failures abound, but the solution is always the same. Whenever possible, orient the design so that the maximum stresses coincide with the direction giving the best mechanical properties. When this is not possible, reduce the design stress so that it corresponds with the properties actually expected.

Service Environment

With the exception of fatigue, more premature product failures have undoubtedly occurred as a result of the service environment than any other single cause. Furthermore, it is very difficult at times to distinguish the failures caused by fatigue from those caused by the service environment. In most cases, these failures have resulted from the failure of the designer to anticipate accurately the service conditions as regards corrosion, wear, or vibration. We will discuss in Chapter 6 several aspects of these topics plus fatigue which have an important influence on design, together with the factors which aggravate or minimize these problems.

Inadequate Materials Specifications

The ultimate results of the materials engineers' input to the design will be the incorporation of the material requirements into the component drawing, i.e., the materials specification. While many large companies have prepared their own specifications for internal use, most companies use the specifications and standards formulated and agreed upon by national organizations such as The American Society for Testing and Materials (ASTM), The American Iron and Steel Institute (AISI), The American National Standards Institute (ANSI), The American Society of Mechanical Engineers (ASME), and The American Society for Nondestructive Testing (ASNT). The basis for these standards is the consensus arrived at by the organization subcommittees. Ordinarily, this agreement is the result of compromise between conflicting views and opinions regarding what is desirable and acceptable to the industrial representatives. The final standard represents that level of capability which the committee believes is the minimum acceptable level for the industry as a whole.

Specific parameters such as chemical composition, mechanical or physical properties, and dimensions will usually be expressed as a range or will be accompanied by a tolerance. It is important in citing the desired specification that the design be capable of functioning properly with the minimum value which could be accepted under that specification.

Though it may appear foolish or unnecessary to make this comment, it is imperative for the designer to know and understand what parameters are controlled by a certain specification. Too often a designer will cite a specification, believing that it guarantees him desirable material without realizing that several specifications may actually be required, each dealing with a particular area of concern.

For example, the engineering drawing for a sand cast aluminum component might cite ASTM B26-72, "Standard Specification for Aluminum Alloy Sand Castings." This would specify chemical composition and mechanical property requirements but would do nothing to control the degree of porosity or shrinkage permitted in the casting. To regulate these factors, additional specifications must be cited, covering, for example, the manner of inspection for such defects and their acceptable levels.

Residual Stresses

Residual tensile stresses are among the most insidious causes of product failures. This topic will be discussed more fully in Chapters 7 and 8. These stresses can contribute to fatigue crack initiation and propagation, stress corrosion cracking, and hydrogen cracking. Residual stresses usually result from fabrication processes such as cold forming, welding, electroplating, machining, and heat treatment. The treacherous nature of such stresses arises from the fact that their presence and magnitude are very often unpredictable. Designs that require fabrication processes which are known to generate residual stress conditions should include provision for stress-relieving treatments such as thermal treatment, shot peening or glass bead peening, vibrating stress relief, etc.

Alternatively, the designer may take advantage of certain processes that induce residual compressive stresses in components. Such procedures include swaging and hydraulic autofrettage,[33] shrink-fit inserts, wrapping,[34] and case hardening.

Inadequate Fracture Toughness

The many advances recently made in the field of fracture mechanics have provided design engineers with additional methods and criteria for predicting the life and the damage tolerance of a structure under certain operating conditions. Since the use of fracture toughness as a design parameter is universally gaining acceptance, designers, particularly those involved in products of a critical nature, should be familiar with these concepts. The interested reader will find some of these concepts summarized by several authors.[35-38]

Inherently, all engineering materials contain flaws. These defects may originate from a variety of sources, such as the melting operation, solidification of an alloy, the mechanical working or shaping process, heat treatment, electroplating, machining, etc., and they range in size from an atomistic scale up to easily observable cracks and defects. Since many such flaws are not easily detectable and most certainly exist, it is beneficial to designers to have a method or criterion that takes flaw size, as well as stress level, into consideration during the design analysis stage.

What is fracture toughness? For the uninitiated, the term fracture toughness implies a parameter or property of a material to withstand failure under applied stresses, in the presence of a flaw. In other words, if a flaw exists in a structure that is subjected to an applied load, at what load does that flaw become critical and result in immediate failure of the structure? This point is a measure of that material's fracture resistance or fracture toughness.

Basically, laboratory-sized specimens are tested to determine the fracture toughness of a material. Then this information is applied to the design or structure taking into consideration the applied stress level, a known or predictable flaw size, and the geometry of the component.

Some of the commercial methods of flaw detection (radiography, ultrasonics, magnetic particle, dye penetrant, and eddy current) are capable of revealing both surface and internal defects down to a level of about 1×10^{-5} to 5×10^{-7} meters.[39] With this capability and fracture toughness data for a material, the designer can predict more reliable stress levels for a structure. In certain cases, structures are designed to carry loads that will likely initiate cracks particularly in the presence of pre-existing flaws or stress concentrations.[40] The designer should anticipate such cracking and realize that under these conditions the component will have a limited lifetime. Again, fracture mechanics concepts can be utilized with laboratory-generated crack growth information to estimate the useful or safe life of the structure.

Many serious failures occur because fracture mechanics concepts and fracture toughness are not well understood and therefore are ignored by designers. The American Society for Testing and Materials has a committee engaged in exploring fracture toughness testing and its application to engineering and design.

A recent consequence of the availability of fracture mechanics to designers is the concept of "fail-safety and damage tolerance." According to this criterion, safety requires that a design will not fail catastrophically even though it contains a crack or has suffered failure of one or more integral parts. It also requires that the damage must be detectable before it reaches a critical stage. Under these conditions, the design is considered fail-safe. This is an extremely valuable tool for designers and engineers and has particular application in critical areas such as the aircraft and nuclear industry.

In summary, it is important to point out that it may no longer be sufficient to consider just the more familiar material properties such as hardness, tensile strength, impact toughness, reduction of area, and elongation. Fracture toughness and crack growth data are rapidly becoming available in the literature and the means for generating such information for a material is also currently well established.

UTILITY AND FUNCTION

In addition to the design factors we have already discussed in this chapter, *product utility* and *product function* should be considered from the liability standpoint at this stage. Nibel and Draper[41] have defined these factors as follows:

Utility — "A product is useful if it functions as planned and is not injurious to morale, good health or good order."

Function — "Sound functional design assures that a product will satisfactorily operate for a reasonable period of time in the manner intended."

These factors not only complement the other phases of design but give the designer an opportunity to further analyze possible hazards in the design or product. He may ask: Do the utility and function of the product provide a desirable or necessary commodity to the public? Does the product fulfill a need or benefit the general welfare? Presumably, the answers to these questions should be affirmative. Then these traits can be weighed against the potential hazards and risks a design or a product may present to consumers.

If possible hazards are discovered and considered at this beginning stage of the manufacturing cycle, and the usefulness or need for the product appears to balance the risk of harm, the designer must examine ways to alleviate these risks or hazards. As we mentioned in Chapter 4, a thorough documentation of the procedures used to balance the risks and utility of a product can be very beneficial to a manufacturer, especially if product liability becomes an issue in the future.

Labels and Warnings

Weinstein et al.[42] point out that there is a growing trend in products liability law to try cases on the basis of "failure to warn."

They state the remedy simply enough: when in doubt, use a warning. Safety warnings are relatively inexpensive and should be used to address risks that cannot, for reasons of cost or usefulness, be eliminated from a product.

Plainly, the incorporation of labels and warnings is a consideration to be addressed in the design stage. Perhaps the design review is an appropriate time. Such warnings can be effective in reducing product liability; the earlier they are established, the better. They leave something to be desired if they are added after problems have occurred in use. Furthermore, warnings and labels should not be considered if a hazard can be economically eliminated from a component by a design change. Such an argument would certainly be deliberated by a jury in a product liability suit. Could the harmful condition have actually been removed by modifying the design? What were the alternatives? What would they cost? Would the alternatives themselves produce another hazardous condition?

We have previously discussed balancing product utility or usefulness against risks. A similar technique would very likely be used in the deliberation of a product liability case, particularly if a known hazardous condition existed in a product and was simply warned against.

The message is clear: warnings and labels are useful and effective in many instances. However, they are not a panacea. They should be used only when no other reasonable means exists for reducing the risks associated with a product or design.

Another aspect of this area of product liability is the use of contradictory warnings. By this term we mean a warning statement that restricts or prohibits a product from being used in a typical or reasonable manner. For example, a recent article[43] involving warnings on automotive jack stands disclosed that certain manufacturers warned against using the jack stands during any under-vehicle repairs or service. Since jack stands are typically utilized to maintain a vehicle in an elevated position so that repairs may be conducted under and about the vehicle, warning against such use is evidently an attempt by the manufacturers to limit their liability in the event of an accident. Will such a maneuver work? The question remains to be settled, but such an obvious contradiction to the implied or suggested use of a product will constitute a weak defense in the event that a user is injured.

As the article referred to above concludes, if this were an acceptable technique of limiting product liability exposure, we would soon see a plethora of warning statements such as:

> "Don't stand on this ladder."
> "Don't drive nails with this hammer."

Obviously, adherence to these warnings would eliminate the hazard but it would also eliminate the function of the device, thus rendering it valueless as a useful tool.

Certain warning and safety information on consumer products is required by law under the following legislation: the Federal Hazardous Substances Act (including the Child Protection and Toy Safety Act), the Poison Prevention Packaging Act, the Flammable Fabrics Act, and the Refrigerator Safety Act. (These acts are outlined in Appendix 1.)

Generally, labeling serves to warn consumers or users at the time they purchase a product that certain hazards are involved.[44] The warnings and precautions on such labels may indicate to a consumer or user that alternative products may be available which are safer to use. If a consumer uses a product which presents a hazard, the label should provide explicit directions for the safe use of the product. Furthermore, particularly in the case of an obviously risky product, labels should indicate remedial action in the event of accident or injury. This type of labeling is typical for poisonous substances.

Labeling should be responsive to the potential consumer or persons who may come in contact with a product. The message or warning must be easily understood. A good example is the recent warning labels included on some poisonous substances which may be accidentally ingested by small children. Instead of, or in addition to, the time-honored skull and crossbones most adults are familiar with, a new logo called "Mr. Yuk" has been introduced by the Pittsburgh Poison Center. This symbol (Fig. 5-19), usually printed in a bilious color, indicates an untasteful or unpleasant substance and apparently is more meaningful to young children than the traditional poison warning label.

Misuse and Abuse

Actually, the designer has to go one step further than designing a product to be functional and safe; he must also design against *foreseeable misuse*. This process can be conducted by the individual designer during the various stages of design or by a review of the design concept before production begins. Early in Chapter 4, we suggested a line of questioning regarding the review of a design. One of the

FIG. 5-19. Poison warning symbol — "Mr. Yuk." (Courtesy of Children's Hospital of Pittsburgh)

questions was, what hazards may be incurred by improper use of the product? Depending on the particular product or design involved, it may be possible to anticipate a number of improper uses and perhaps abuses. These conditions should be thoroughly explored and considered as potential product liability problems.

Let us examine several examples where misuse has been anticipated and the manufacturers have attempted to eliminate a hazard resulting from misuse. For example, electrical equipment can be inadvertently overloaded by careless users. In essence, this amounts to misuse of the circuitry, but it is also a foreseeable misuse and one that can be designed against. Depending on their wire size, circuits should be protected from overloading by proper-size fusing or circuit breakers. Similarly, electrical power tools are frequently misused and abused, thus creating a shock hazard to the user. Manufacturers have attempted to design against this by providing double insulation on certain power tools and by using plastics, instead of metals which have considerably higher electrical conductivity.

More recently a device called a ground fault interrupter (GFI) has become commercially available and is gaining in popularity.[45] The

GFI can be installed in a circuit panel in place of a normal circuit breaker, or a slightly different version can actually be plugged in at the work site. This device senses any current to ground and opens the circuit in a fraction of a second, thereby preventing serious shocks.

These design techniques can prevent electrical shocks to the consumer under a wide variety of applications and operating conditions. The result is a reduction in user-related injuries and a corresponding decrease in product liability.

The use of a carpentry hammer (claw hammer) for any purpose other than driving or pulling carpentry nails actually constitutes a misuse of that product. But how many different uses does this type of hammer really see? The answer, as practically everyone knows, is literally dozens of other uses, constructive and otherwise. Our point is that it is common knowledge that users misuse claw hammers and occasionally this treatment results in an injury to the user. For instance, chipping of the striking head can occur when a carpentry hammer is used to hammer very hard objects or materials. A hammer which has undergone just such damage is shown in Fig. 5-20. As a result of such misuse, fragments have struck users in the face and eyes, inflicting serious injuries.

It is possible to rim temper the edge of the striking surface, thereby reducing its hardness and its tendency to spall, and certain manufacturers have done so. However, this condition cannot be completely designed out of the hammer without seriously affecting its intended function, which of course is to drive nails. Therefore, manufacturers display warning labels on their hammers, cautioning against misuse of the product, and instructing the operator in some safeguards involving the correct use of the hammer.

However, the best warnings in the world will not prevent product liability. In fact, defect-free parts are occasionally involved in product liability litigation. Take the example of the carpentry hammer just discussed. Such a tool may have been designed properly and manufactured correctly, containing no defects. When originally purchased, the hammer very likely had a warning label prominently displayed. Yet, through misuse, perhaps striking a hard object such as a steel wedge or chisel, a person receives an eye injury from a hammer fragment. Is the hammer manufacturer liable? Our experience is that rather than argue the merits of the case in court, the manufacturer often opts for settlement before the case reaches trial. So, liable or not, the costs of a product liability litigation are sustained by industry and eventually the consumers.

FIG. 5-20. Overall view of carpentry hammer, and close-up of striking face showing chipped edges.

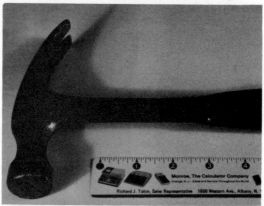

Safety Features and Devices

The willingness of courts to accept frailties in human behavior and understanding is the reason why manufacturers have been unsuccessful in persuading the courts that liability should be limited to the intended use of their products.[45] In other words, the focus has been placed on the product no matter who buys it or eventually uses it. The result is that the courts are favoring the position that safety devices and guards should be built into the basic product, not available as optional equipment.

Earlier we stated that for safe and reliable operation, hazards should be eliminated or minimized in the design stage, perhaps during the design review. However, if the hazards are not eliminated for some reason, safety devices or guards are a must. Furthermore, the eventual users of such equipment must be educated or trained in the proper use of such devices.

Certain equipment *must* be guarded according to OSHA regulations (Title 29, Code of Federal Regulations, Part 1910 — Safety and Health Standards for General Industry), especially machinery and power tools. Guarding requirements differ in their particulars according to the machine, but all machines are covered under the general clauses [1919.212 (2)(1)]:

> One or more methods of machine guarding shall be provided to protect the operator and other employees in the machine area from hazards such as those created by nip points, rotating parts, flying chips, and sparks. Examples of guarding methods are: barrier guards, two-hand tripping devices, electronic safety devices, etc.

Although manufacturers are incorporating some safety features in their products, many devices are left as options. Rotary lawn mowers are being produced with deflector shields over the discharge chute, but few contain shields at the rear of the mowing deck. Certain brands of woodcutting chain saws contain chain brakes (deadman's switch) and safety tips to prevent kickback, but not all manufacturers use such devices, and some offer them as optional extra equipment.

Certainly this extra equipment poses an economic question to a manufacturer, as well as a technological problem to the engineering and design personnel. The increased costs must be passed along, in part, to the consumer, who may or may not appreciate the safeguards. Therefore, if a safety device is definitely warranted, it is up to the designers and engineers to produce an adequate mechanism at the least cost. Remember, "First cost is not necessarily final cost." The additional cost of the device or feature must be balanced against the potential risk of product liability suits and insurance. We have cited numerous examples in earlier chapters which have borne out the cliché "An ounce of prevention is worth a pound of cure."

UNFORESEEN SERVICE CONDITIONS

A final note regarding product design and its liability implications. There are occasional instances where a product is not abused or misused but simply fails because of unforeseen operating or environmental conditions. If such a situation is the act of a higher authority than man, then perhaps there is no possible earthly liability. However, if the unforeseen service condition is not an act of God, it may have material or design implications. A striking example of unforeseen or unanticipated service conditions is the flammability of certain materials in the following categories:

Textiles — used for clothing, bedding, and furnishings.[47-49]
Plastic foams — used for furniture, furnishings, and insulation.[50]
Aerosols — used for hair sprays, insect repellents, etc.

The large-scale use of the flame-retardant compound Tris (tris 2, 3, dibromopropyl phosphate) was initiated in 1973 to satisfy the requirements for flammability standards in infants' and children's sleepwear. Subsequently it was found that Tris is mutagenic, and the sale of garments treated with Tris was banned in April 1977.[51] Not only did this situation leave manufacturers, distributors, and retailers stuck with an estimated $125 million worth of Tris-treated garments but also raised the specter of future legal problems associated with people who wore Tris-treated clothing and have developed cancer.

SUMMARY

The prudent design of an engineering component or structure involves not only *mechanical* and *geometric design*, but also *process design*. Cooperation between the mechanical designers and the process engineers is absolutely necessary for the best possible performance and reliability of any product. Concomitantly, this approach will tend to minimize the product liability potential of the producer.

Although the details of many manufacturing processes and their relation to product liability will be discussed in Chapters 7 and 8 it is appropriate at this point to ask, what do we actually mean by the term process design? For the purposes of this discussion, we simply mean, "the technique of planning the proper and correct treatments in the right sequence, which will produce the required or desired properties in a part or material." The following categories may help to explain this procedure:

Design of Alloys and Melting: What chemical composition best suits the intended product from the standpoint of mechanical properties, formability, etc.? Which melting method or combination of processes will produce an acceptable and reliable material at least cost?

Design of Forming: Which fabrication methods are suitable to produce a particular component or shape? How much reduction or deformation is acceptable?

Design of Heat Treatment: A thorough consideration must be made of the factors which will affect the results of heat treatment and produce an acceptable material with the required properties.

It should be evident that process design is just as intricate a procedure as the mechanical design of a part. Also, the aspects of mechanical design discussed in this chapter — materials selection, specifications, the utility and functioning of a component, etc. — cannot be separated from process design. If such a separation of design categories exists or is produced at some stage of manufacturing, the potential for product liability is increased. As the design stage moves from the conceptual phase to the fabrication phase, the interaction of many departments and skills is important. Such interaction is depicted by the chart in Fig. 4-4. Hopefully, the well-known adage "One hand doesn't know what the other is doing" does not apply, for such a situation invites future problems in product performance, reliability, and liability. Appropriately, in their chapter on product defects, Weinstein et al.[52] state that "the design flaw is what exists in all products of that make or of that kind." This definition encompasses errors in process design which result in a defective product, because a flaw in process design can just as easily turn out defective part after part as a flaw in mechanical design.

The following chapters on service conditions and materials processing factors may help to define some of the hazards and pitfalls that the designer and engineer can encounter in the utilization of structural and engineering materials.

REFERENCES

1. "Structural Failures: Modes, Causes, Responsibilities," ASCE National Meeting on Structural Engineering, Cleveland, Ohio, April 1972, American Society of Civil Engineers, New York, 1973.
2. "An Inquest Into Why All the Roofs Fell In," *Business Week*, Feb. 6, 1978, p. 46.
3. Griffel, W., *Handbook of Formulas for Stress and Strain*, Frederick Ungar Publishing Co., New York, 1966.
4. *Handbook of Experimental Stress Analysis*, M. Hetenyi, Ed., John Wiley and Sons, New York, 1950.

5. *Standard Handbook for Mechanical Engineers*, T. Baumeister and L. S. Marks, Ed., McGraw-Hill, New York, 1958.
6. *Mathematical Handbook for Scientists and Engineers*, G. A. Korn and T. M. Korn, Ed., McGraw-Hill, New York, 1961.
7. *Handbook of Engineering Mechanics*, W. Flugge, Ed., McGraw-Hill, New York, 1962.
8. Huebner, K. H., *The Finite Element Method for Engineers*, John Wiley and Sons, New York, 1975.
9. Zienkiewicz, O. C., *The Finite Element Method in Engineering Science*, McGraw-Hill, London, 1971.
10. Krouse, J. K., "Finite-Element Update," *Machine Design*, Jan. 12, 1978, pp. 98-103.
11. *Interactive Computer Graphics in Engineering*, L. E. Hulbert, Ed., American Society of Mechanical Engineers (ASME), New York, 1977.
12. Dieter, G. E., Jr., *Mechanical Metallurgy*, McGraw-Hill, New York, 1961.
13. Ref 5, p. 5-29.
14. Neugebauer, G. H., *Product Eng.*, Vol. 14, 1943, pp. 82-87.
15. Chow, W., *Cost Reduction in Product Design*, Van Nostrand Reinhold Co., New York, 1978, p. 209.
16. Ibid., p. 210.
17. Timoshenko, S., *Strength of Materials, Part II — Advanced Theory and Problems*, D. Van Nostrand Co., Inc., Princeton, N. J., 1956, p. 145-146.
18. Singer F. L., *Strength of Materials*, Harper and Row, New York, 1962, p. 567.
19. *Wood Handbook*, Forest Products Laboratory, U. S. Dept. of Agriculture, 1940, pp. 50-53.
20. Chow, Ref 15, p. 171.
21. Ibid., p. 172.
22. Marshall, T., *Product Eng.*, Nov. 17, 1969, pp. 72-74.
23. "Fastening and Joining — 1978 Reference Issue," *Machine Design*, Nov. 16, 1978.
24. Laughner, V. H., and Hargan, A. D., *Handbook of Fastening and Joining of Metal Parts*, McGraw-Hill, New York, 1956, p. 204.
25. Nibel, B. W., and Draper, A. B., *Product Design and Process Engineering*, McGraw-Hill, New York, 1974, p. 742.
26. Chow, Ref 15, p. 306.
27. Colangelo, V. J., and Thornton, P. A., "Stress Corrosion Cracking in Low Alloy Steel Spindles," in *Microstructure Science*, Vol. 7, Elsevier North Holland, Inc., New York, 1979, p. 51.
28. *Metals Handbook*, 8th Ed., Vol. 10, "Failure Analysis and Prevention," American Society for Metals, Metals Park, Ohio 44073, p. 470.
29. Nibel and Draper, Ref 25, p. 743.
30. Ibid., p. 744.
31. Annual Book of ASTM Standards, E-399, "Plane Strain Fracture Toughness Testing of Metallic Materials," American Society for Testing and Materials, 1916 Race St., Philadelphia, Pa. 19103.
32. *Business Week*, Sept. 22, 1973, p. 32.
33. Davidson, T. E., Kendall, D. P., and Reiner, A. N., "Residual Stresses in Thick-Walled Cylinders Resulting From Mechanically Induced Overstrain," Paper No. 762, SESA Spring Meeting, Seattle, Washington, May 8-10, 1963.

34. Davidson, T. E., and Kendall, D. P., "The Design of Pressure Vessels for Very High Pressure Operation," Watervliet Arsenal Technical Report, WVT-6917, 1969.

35. Tetelman, A. S., and McEvily, A. J., Jr., *Fracture of Structural Materials*, John Wiley and Sons, New York, 1967.

36. "Fracture Toughness Testing and Its Applications," ASTM STP-381, American Society for Testing and Materials, 1916 Race St., Philadelphia, Pa. 19103 (1965).

37. "Application of Fracture Toughness Parameters to Structural Metals," Metallurgical Society Conf. (AIME), Vol. 31, Gordon and Breach Science Publishers, New York, 1966.

38. Brown, W. F., Jr., and Srawbey, J. E., "Plane Strain Crack Toughness Testing of High Strength Metallic Materials," ASTM STP-410, American Society for Testing and Materials, 1916 Race St., Philadelphia, Pa. 19103 (1966).

39. Vary, A., "Nondestructive Evaluation Technique Guide," NASA-SP-3079, National Aeronautics and Space Administration, Washington, D.C., 1973.

40. Broek, D., *Elementary Engineering Fracture Mechanics*, Noordhoff International Publishing, Leyden, The Netherlands, 1974, p. 7.

41. Nibel and Draper, Ref 25, p. 20.

42. Weinstein, A. S., et al., *Products Liability and the Reasonably Safe Product*, John Wiley and Sons, New York, 1978, p. 62.

43. *Consumer Reports*, June 1980, p. 378.

44. Hicky, L. E., *Product Labeling and the Law*, American Management Association, New York, 1974, p. 24.

45. "Protection From Electrical Shocks," *Consumer Reports*, Feb. 1979, p. 117.

46. Weinstein et al., Ref 41, p. 137.

47. *Consumer Reports*, Apr. 1971, p. 238.

48. Young, C., "Burn-Resistant Fabrics: There's Hope Ahead," *FDA Papers*, July/Aug. 1971, GPO: 1971-482-085/4; Mathers, W., and Klam, K., "When Clothing Ignites," *FDA Consumer*, Oct. 1972, GPO: 1973-515-080/59.

49. 6th Annual Report of the U.S. Consumer Product Safety Commission on Flammable Fabrics Data (1978), Jan. 1979, U.S. Consumer Product Safety Commission, Washington, D.C. 20207.

50. Schafran, E., "Development of Flammability Specifications for Furnishings," *Fire Journal*, Mar. 1974, p. 36.

51. Abelson, P. H., "The Tris Controversy," Editorial, *Science*, Vol. 197, No. 4299, July 8, 1977, p. 113.

52. Weinstein et al., Ref 41, p. 29.

6

Materials Factors That Affect Product Liability

In this chapter, we shall examine several of the more common failure modes with an eye toward the relation of these modes to potential liability in the event of a product failure. This chapter is not intended to encompass all the information currently known about a particular failure mechanism but rather to familiarize the reader with potential problem areas and to show that failure mechanisms are very often complex, with one mechanism responsible for crack initiation and another for propagation.

DUCTILE-BRITTLE BEHAVIOR

Ductile Fractures

Ductile failures occur when the material in a component is over-stressed. Because of this, ductile fractures are high energy fractures. They are characterized by stable crack propagation. If the load causing the fracture is removed, crack propagation ceases. Ductile fractures exhibit evidence of plastic deformation. On a macroscopic scale, one can usually find shear lips or other evidence of plasticity, while on a microscopic scale the fracture will show dimpling.

As previously stated, ductile fractures occur when the applied load is above the yield strength of the material. This may be the result of several factors:

(a) The yield strength of the material was not as high as required by the design. This may result if errors in procurement occur, or if the material is substandard.

(b) The normal service conditions place loads on the system which exceed those anticipated by the designer. This type of failure is directly attributable to the designer and represents a serious failure of the design process.

(c) Abnormal loading places critical stresses in the system. This may be a nonforeseeable occurrence or it may be related to a failure in the design process. What constitutes abnormal usage is not cut and dried. Judgment is often required to separate traumatic effects from those occurring as a result of failure of the design process. The plastic deformation of an automobile frame resulting from a collision is a clear example of abnormal loading. So is the rupture of a water pipe because of freezing, as shown in Fig. 6-1. On the other hand, the collapse of a roof structure may be the result of an abnormal snow load or it may result from fabrication defects or improper design. Thus, the

FIG. 6-1. Ductile failure of a water pipe due to internal pressure resulting from a freeze.

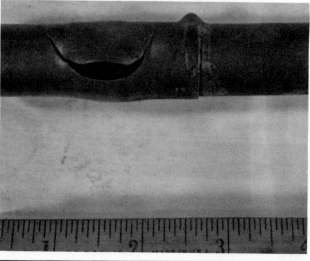

situation is complex and requires careful analysis to determine the true cause.

Brittle Fractures

Brittle fractures occur at stresses far lower than the yield strength of the material under stress. These fractures are usually associated with cracks or other flaws in the structure and are characterized by low energy absorption and a lack of plastic deformation in the gross appearance. On a macroscopic scale, the fractures tend to be flat and do not exhibit shear lips or other evidence of plasticity. On a microscopic scale, the fracture surface usually exhibits the morphology associated with cleavage, quasi-cleavage, or intergranular fracture (Fig. 6-2).

The theory behind brittle fracture is based largely upon the work of Griffith, who did much of the initial work in quantitatively measuring fracture strength. His theory considered the energy required to fracture a material with a penny-shaped crack oriented perpendicular to the applied stress. He developed an energy balance with three main components: the strain energy in a system without a void, surface energy required by the creation of fracture surfaces, and the energy released by a moving crack.

As a crack moves, strain energy is released but surface energy is consumed. If there is a balance, the crack is stable and no propagation occurs. If there is an imbalance, the crack is unstable and the crack grows. The Griffith concept considers only the energy required for brittle fracture with no consideration of the energy absorbed in plastic deformation at the crack tip. The consideration of this latter factor by Orowan[1] and Irwin[2] set the stage for the development of fracture mechanics. The fracture mechanics approach considers the effect of a crack upon the actual stresses on the crack tip. This stress intensification is expressed as a stress intensity factor (K). When the stress intensity reaches a critical value (K_{IC}), fracture occurs. The K value for a particular specimen at any point in time depends upon the crack length, the geometry of the specimen, and the stress level.

Fracture toughness testing is usually conducted on specimens of relatively simple but standard configurations. However, once this data is obtained, it can be transferred to a more complex component if the K calibration has been computed for that configuration. Much of the work currently under way in the fracture mechanics area[3] involves the utilization of these concepts in the design stage to prevent brittle fracture and to control crack growth.

FIG. 6-2. Transmission electron microscope fractograph showing (a) cleavage, (b) quasi-cleavage, (c) intergranular fracture (6000×). Courtesy of L. McNamara)

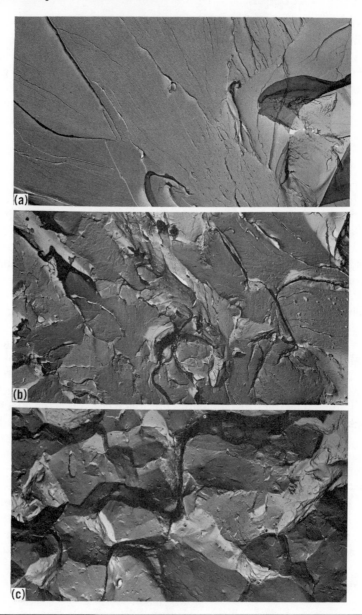

Does all this sophisticated knowledge mean that brittle fractures are a thing of the past? It does not appear so. In the late 1960's, it became apparent that a rash of injuries and fatalities in automobile accidents in New York State were associated with the failure of bridge railings.[4] Subsequent investigation revealed that the railings were failing under moderate impact, as a result of fracture of the cast aluminum vertical posts, thereby permitting the vehicles to fall from the bridges. Upon further investigation, it was disclosed that impact strength of the alloy had not even been a design consideration, that the design was based only upon the obtainable yield strength of the aluminum alloy. Ultimately, as a result of this design error, thousands of feet of railing had to be replaced in order to assure the safety of the public.

Ductile-to-Brittle Transition

Unfortunately for the design engineer, alloys which exhibit ductile behavior at one temperature may display brittle behavior at a lower temperature. This phenomenon, called the ductile-to-brittle transition, represents a serious design consideration. Figure 6-3 schematically illustrates this behavior as it occurs with the Charpy

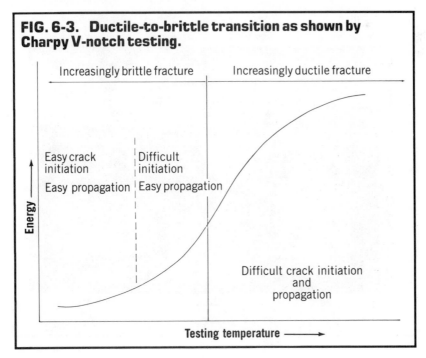

FIG. 6-3. Ductile-to-brittle transition as shown by Charpy V-notch testing.

Increasingly brittle fracture   Increasingly ductile fracture

Energy →

Easy crack initiation
Easy propagation

Difficult initiation
Easy propagation

Difficult crack initiation
and
propagation

Testing temperature ——→

V-notch impact test. As çan be seen, there is a decrease in the impact strength from the upper shelf obtained at higher temperatures through a transition to a lower range of impact strengths at lower testing temperatures. This loss in impact strength is also characterized by a change in the fracture appearance, as shown in Fig. 6-4.

As the testing temperature decreases, the energy level decreases. The percentage of lateral expansion (or increase in the width of the bar), which is a measure of the plasticity, also decreases. The fracture appearance also changes, with the percentage of fibrous fracture increasing as the testing temperature increases.

Each of these observed phenomena may be the criterion used to establish the "transition temperature," i.e., the testing temperature at which the behavior changes from ductile to brittle. However, this transition temperature is usually set on an energy basis, e.g., 50% of the upper shelf value, or 15 ft-lb, or on a fracture appearance transition temperature (FATT), depending upon which value is the most conservative for the alloy in question. Regardless of the method

FIG. 6-4. Change in fracture appearance and in lateral expansion as a result of testing temperature (2×).

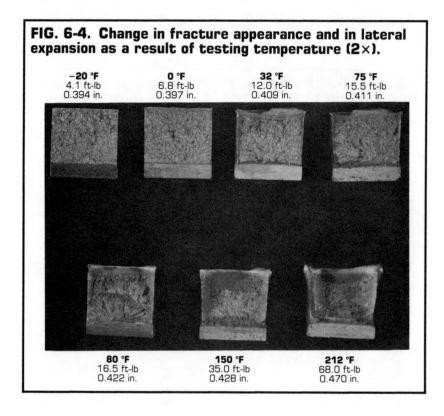

−20 °F	0 °F	32 °F	75 °F
4.1 ft-lb	6.8 ft-lb	12.0 ft-lb	15.5 ft-lb
0.394 in.	0.397 in.	0.409 in.	0.411 in.

80 °F	150 °F	212 °F
16.5 ft-lb	35.0 ft-lb	68.0 ft-lb
0.422 in.	0.428 in.	0.470 in.

chosen to establish the transition temperature, it is imperative that this transition temperature be below the service temperature for the alloy for any design in which low temperature service is anticipated. Figure 6-5 demonstrates the behavior of two alloys with differing transition temperatures based upon a criterion of 15 ft-lb. An examination of this diagram discloses that alloy A is the proper choice for the design based upon the minimum service temperature. The use of alloy B would create, under certain service conditions, an undesirable situation in which the structure would have severely impaired impact resistance.

This, of course, is so obvious that one may question why it would ever be a problem. One way the problem arises is when the selected alloy, as a result either of improper heat treatment and processing or of variations or impurities in the chemical analysis, does not yield the typical results expected of the alloy. This is one area where in-house testing of the raw material can prove extremely beneficial for two reasons: first, because it saves expenditures involved in processing inferior material; second, and most important, because it prevents a potentially dangerous product from entering the distribution system.

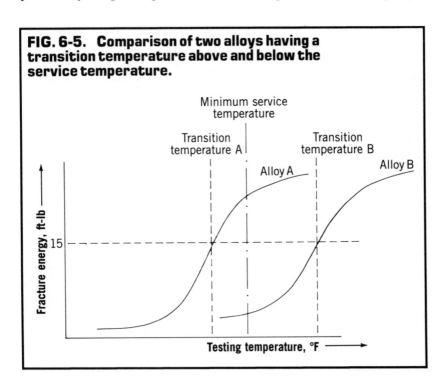

FIG. 6-5.   Comparison of two alloys having a transition temperature above and below the service temperature.

Another way the ductile-to-brittle transition temperature can become a design problem is when the designer fails to anticipate the lowest temperature which can be encountered in service. There is no easy solution here except that the designer must be thoroughly familiar with the possible application and provide sufficient latitude in the design temperatures to accommodate possible deviations. Otherwise he risks the potential catastrophic failure of the system if lower than usual temperatures are encountered.

FATIGUE

Background

Fatigue failures are those which occur as the result of repeated, i.e., cyclic, loads. These fractures are characterized by three stages: crack initiation, crack propagation, and fast fracture. Although these phases are controlled by differing mechanisms, they are not completely separate and some overlap occurs.

Typically, a fatigue fracture will occur as follows: Under cyclic loading a crack will initiate at a region of high localized stress — at the root of a notch, for example. With continued cyclic loading, crack growth progresses until at some point, depending upon the material, the stress level, and the design, the critical crack depth is reached and catastrophic fracture occurs.

This does not mean that in the absence of a stress raiser fatigue will not occur and the component will last indefinitely. With sufficiently high stresses, even unflawed specimens will also fail as a result of fatigue if cycled long enough. All factors considered, the higher the stress level, the lower the number of total cycles required for failure.

What do we mean by a stress cycle? This is a single segment of the stress-time function which is periodically repeated. Figure 6-6 illustrates reversed and fluctuating stress cycles commonly used in fatigue testing and indicates schematically many of the terms used. In the real world, the variation of stresses is much more complex; modern fatigue tests have attempted to simulate the effects of variable cycles through the use of complex loading programs in which the load spectrum is varied. Figure 6-7, taken from Buxbaum,[5] shows the kinds of load spectra resulting from various applications; Fig. 6-8 shows the use of block programming[6] to simulate various load spectra.

Since fatigue data generally show large scatter, they may be presented in several ways depending on the interpretation and the manner in which they will be used. Most curves presented by a material producer are so-called typical curves, usually the average curve drawn by eye through the fracture points on an S-N diagram. This typical curve is not safe for design purposes since obviously about 50% of the material will fall below the curve. Sometimes the low boundary of the fracture points of a series of tests is given, but even this is not safe: usually only a few specimens are tested and there is the probability that if many were tested, an unknown quantity would fail below the low boundary formed by the few points.

In order to present more conservative data for design purposes, statistical and probability methods are often used. One method is as follows:

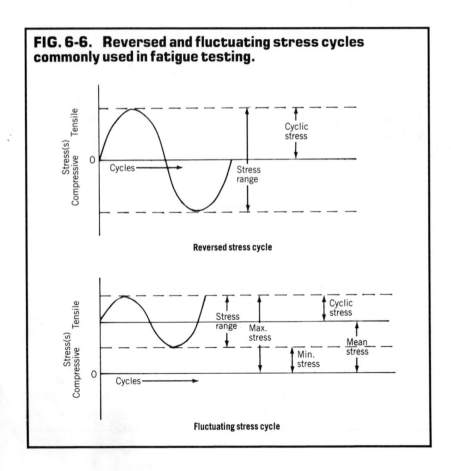

FIG. 6-6. Reversed and fluctuating stress cycles commonly used in fatigue testing.

An average curve is drawn through the test points so that the sum of the ratios of all the test point stresses to the corresponding curve stresses at a certain number of cycles, divided by the number of points, is approximately 1, or:

$$\bar{x} = \frac{x}{n} \simeq 1$$

where $x = \dfrac{\text{test point stress}}{\text{curve stress}} = \dfrac{S_t}{S_b}$ and n = number of test points.

Statistical analysis is then used to establish the standard deviation. The use of standard deviations helps determine the risk or the percentage of specimen test points which will fail within a certain range of values. The values of standard deviations for a large number of normally distributed points are as follows:

Deviation	% of Test Points Within Deviation	% of Test Points Below Minimum Deviation Boundary
$\pm\sigma$	68.2	15.9
$\pm 2\sigma$	95.4	2.3

FIG. 6-7. Typical load-time histories resulting from various applications.

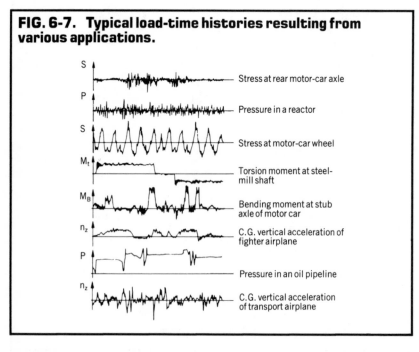

S	Stress at rear motor-car axle
P	Pressure in a reactor
S	Stress at motor-car wheel
M_t	Torsion moment at steel-mill shaft
M_B	Bending moment at stub axle of motor car
n_z	C.G. vertical acceleration of fighter airplane
P	Pressure in an oil pipeline
n_z	C.G. vertical acceleration of transport airplane

FIG. 6-8. Examples of block spectrum loading sequences.

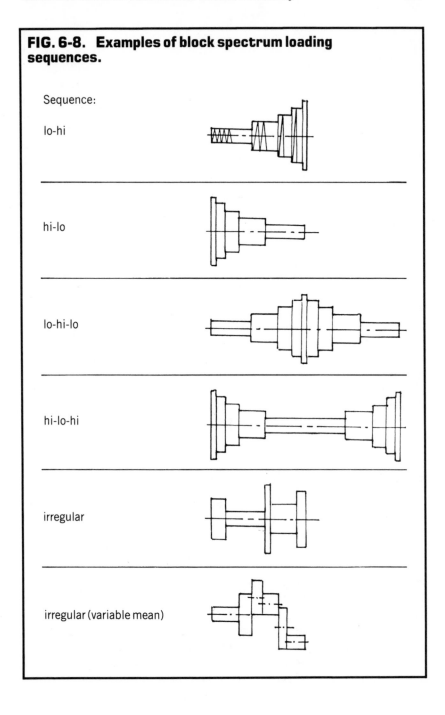

Sequence:

lo-hi

hi-lo

lo-hi-lo

hi-lo-hi

irregular

irregular (variable mean)

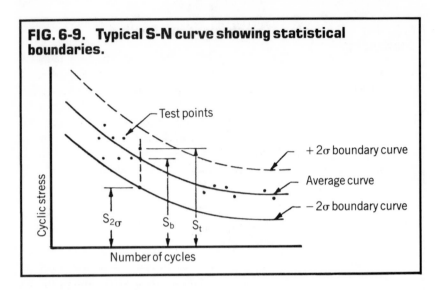

FIG. 6-9. Typical S-N curve showing statistical boundaries.

This means that if boundary curves are drawn corresponding to 2σ above and below the average curve, such boundary curves will include 95.4% of all test points. Thus in a large size sample, 2.3% of the specimens would fail below the -2σ boundary curve. Since in the majority of laboratory tests the sample size is small, the -2σ curve usually falls below all test points; therefore, it is a better estimate of the stress at which there is a uniformly low probability of failure than a curve through the lowest test points. These boundaries are illustrated in Fig. 6-9.

Effects of Stress Concentration

It is generally accepted that the presence of stress raisers facilitates the formation of a crack and thereby shortens the fatigue life of a structure. These stress raisers may result from factors implicit in the design, such as keyways, grooves, and inadequate fillet radii. They may be mechanical and result from the machining or fabrication process. Tool marks and welding cracks would fall into this category. They may also be metallurgical in nature — for example, porosity or inclusions existing within the metallurgical structure. Service effects may also create surface damage by means of pitting or fretting, for example, and thereby facilitate crack formation. Regardless of the precise origin, fatigue fracture illustrates the fact that failures are often complex, with varying causes. The catastrophic occurrence of a fatigue failure caused by a weld defect can be seen in the following example. A fatigue crack severed the main structural member support-ing the passenger seat in an amusement park ride. The location of the failure is indicated in the photo at left in Fig. 6-10. The fatigue crack originated at an undercut adjacent to the weld where the support hanger was welded to the bow. A similar defect is shown on an adjacent support bow in the photo at right in Fig. 6-10. This notch-like defect was created during the welding operation and was not detected during subsequent inspection. Painting masked the under-cut and the unit was placed into service. After approximately two years of use the support bow fractured due to fatigue as shown in Fig. 6-11, dropping and seriously injuring two passengers.

Effect of Surface Conditions

In most service applications, the maximum stress occurs at or near the surface of the component. Therefore, the fatigue life is gener-ally sensitive to the surface condition and is affected by such things as surface finish, chemical and physical characteristics of the surface, and any residual stresses which may be present.

Surface Finish. If one examines the surface of a machined component under high magnification, it is immediately apparent that the surface has been plowed into a regular array of small notches by

---

FIG. 6-10. Fatigue failure in an amusement park ride resulting from a weld defect.

Location of the failure   Type of weld defect (undercut)

the cutting edge (edges) of the tool. As the surface finish of the component becomes coarser, the depth of the notches increases and the traumatization of the surface is similarly increased. This notch effect resulting from coarse surface finishes causes a decrease in the fatigue life. The following study[7] conducted on SAE 3130 alloy illustrates the effect of surface finish:

Machining operation	Surface finish, micro-in.	Fatigue life, cycles(a)
Lathe................................	105	24,000
Partly hand polished..................	6	91,000
Hand polished	5	137,000
Ground..............................	7	217,000
Ground and polished	2	234,000

(a) SAE 3130, complete stress reversal at 95,000 psi. The data also illustrate another point, that the type of machining operation as well as the surface finish affects fatigue life. This apparently relates to the size of the root radius of the notch and the degree of trauma created by the particular machining operation.

Surface Treatments. Surface treatments cause changes in material properties that can significantly affect fatigue life. These effects may be divided into those which decrease life and those which increase life. Electroplating very often decreases fatigue life. For example, when chromium plate is deposited, tensile stresses are developed in the electroplating and in the immediately underlying base metal. These stresses can cause cracking which is, in effect, instantaneous fatigue crack initiation. In addition, in an electroplating bath, hydrogen as well as the deposited metal is liberated at the cathode. When combined with the tensile stresses developed in plating, this can lead to hydrogen embrittlement and a potentially catastrophic service life.

Decarburization of steel is a surface condition which often occurs when heat treating without a protective atmosphere or at moderately high temperature, or at moderately high temperature service conditions. It also occurs during solidification in some casting methods, notably investment casting. This decarburization produces a lower strength alloy at the surface of the structure and reduces fatigue life, since it facilitates the formation of a crack in the lower strength surface layer.

Conversely, carburizing increases fatigue life, as do nitriding and flame and induction surface hardening. This occurs either as a result

FIG. 6-11. Fracture surface of support bow showing beach marks.

of strengthening of the surface material or by the generation of residual compressive stresses or a combination of the two factors.

The relative reduction in fatigue life with various surface finishes and with aggressive environments is also illustrated in Fig. 6-12.[8] It can be seen that the better the surface finish, the higher the fatigue life.

Microstructure

A general study of fatigue life shows that it is affected by the microstructure of the component, that is, it is structure sensitive. Although it is influenced by grain size, the results are variable. For nonferrous materials and annealed steel, the fatigue strength in-

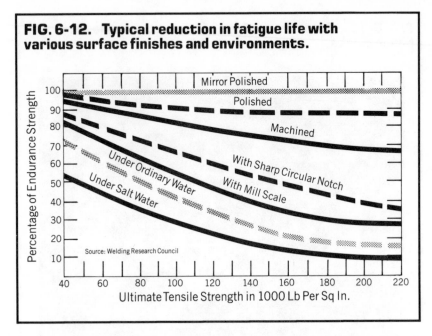

FIG. 6-12. Typical reduction in fatigue life with various surface finishes and environments.

Source: Welding Research Council

creases as grain size decreases. However, in unnotched heat-treated steel, there is no observable effect. Apparently the effects of the microconstituents present override those produced by grain size.

Much of the data published on the effects of microstructure upon fatigue concerns steel. The same strength can be achieved in steel with a variety of microstructures. At a constant strength, the fatigue life of a coarse pearlite is lower than that of a spheroidal microstructure because of the shape of the carbide particles. The rounded carbides in the spheroidized specimen have lower stress-concentrating effects, which result in a longer life. This observation can be generalized to other types of microstructures. The fatigue life of a coarse, angular microstructure is usually lower than that of a fine, rounded microstructure.

In steel, the longest fatigue life is achieved with a tempered martensite microstructure. Slack quenching, which results in a mixed microstructure of martensite and bainite or ferrite, decreases fatigue life. As the tempering temperature decreases and strength increases, the material becomes more sensitive to notches and, therefore, more sensitive to surface conditions. More care must be taken in the design and preparation of the part.

Orientation. Working of metals by rolling or forging causes mechanical fibering, that is, the alignment of grains, chemically segregated areas, and inclusions parallel to the rolling direction. Fatigue life is usually anisotropic; it is lower when tested transverse to the rolling direction. The effect varies with the material and test conditions but becomes more pronounced with increased load, increased yield strength, decreased ductility, decreased cleanliness, and the change from smooth to notched bar. A general relation describing the anisotropy of fatigue life shows that transverse life is approximately 0.6 to 0.7 of the longitudinal life. Although such relations are desirable and sometimes useful, it must be remembered that they are general, and specific cases can vary significantly.

One of the main causes of anisotropy in forged and rolled products is the presence of inclusions and regions of segregation. The volume, shape, and orientation of these inclusions markedly affects fatigue life as well as impact strength. This inclusion effect is one of the prime reasons for the use of special melting techniques such as VAR (vacuum arc remelting) and ESR (electroslag refining). In fact, it has been recognized[9] that the use of special melted steels may significantly affect product liability exposure and this factor is an additional reason for its use.

CORROSION

Products exist and function in various operating environments. Commonly there is some interaction between the material used in the construction of the product and the environment in which it is used. This interaction is known as corrosion. Several types of corrosion have been classified, all of which are capable of causing failure. Among the more common forms are general corrosion, pitting, galvanic corrosion, intergranular attack, and stress corrosion cracking. Types of corrosion most affected by design are crevice corrosion, galvanic corrosion, and stress corrosion cracking. While the other types of corrosion are obviously affected by the environment and choice of material, they are much less affected by the subtleties of design.

Crevice Corrosion

Crevice corrosion is a type of corrosive attack which is associated with the confined spaces or crevices formed by certain mechanical configurations, such as tapped joints, gasket interfaces, tubular sleeves, and so forth.

Localized corrosion occurs in the crevice at a rate many times higher than would be expected from general attack on the unprotected metal. The cause is largely the result of oxygen or metal ion concentration cells. Oxygen concentration cells occur in deep narrow crevices or under loosely adherent deposits in which the diffusion of oxygen is retarded and differences in oxygen concentration result, with attendant differences in electrochemical potential. The regions of low oxygen concentration are anodic relative to the high oxygen regions and therefore corrode more readily. Similar results occur when metallic ion concentration cells exist. The metal in the region of low ion concentration tends to go into solution (corrode).

Crevices can exist in any assembly, but there appears to be a geometrical dimensional requirement in order for the crevice to create corrosion problems. The crevice must be close fitting, with dimensions of only a few thousandths of an inch or less, for corrosion to occur. Although the limits of the gap have not been defined, crevice corrosion does not occur in larger spaces.

Both approximating surfaces need not be metal for crevice corrosion to occur. It has been reported in crevices formed by a number of nonmetallic materials, such as polymers, glasses, and rubber, that are in contact with metal surfaces. This fact is of particular importance in applying and selecting gasket materials.

Several guidelines are helpful in minimizing crevice corrosion problems with a design. While adherence to these guidelines may not eliminate crevice corrosion problems, it will most certainly minimize them:

Keep crevices or confined narrow gaps to a minimum. Where possible, use butt joints rather than bolted or riveted lap joints. Where lap joints cannot be eliminated, use sealants to prevent entry of the corrodent.

Avoid sharp corners, dead ends, and stagnant areas. The design should provide for drainage to prevent the accumulation of moisture and debris.

Do not use absorbent packings and gaskets. Use materials with low wettability.

Minimize the introduction of foreign material into the system and provide for its removal to prevent the formation of sediment deposits. Use continuous welds on lap joints to prevent the ingress of corrodent.

Figure 6-13 illustrates several design concepts which can be applied to help minimize crevice corrosion.

FIG. 6-13. Design recommendations useful in minimizing crevice corrosion.

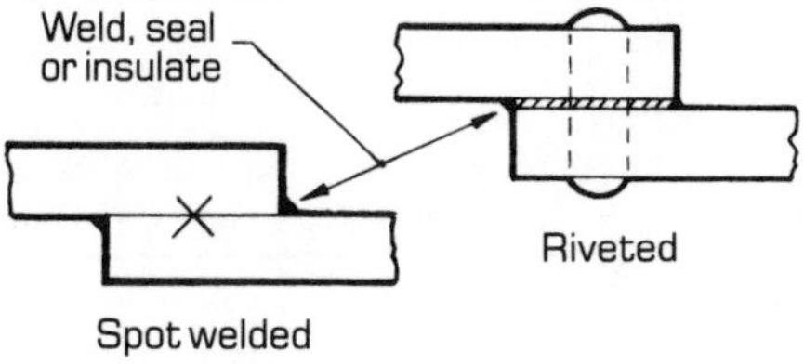

Weld, seal or insulate

Spot welded

Riveted

In lap joints, use of welds, insulating material, or a sealer is recommended.

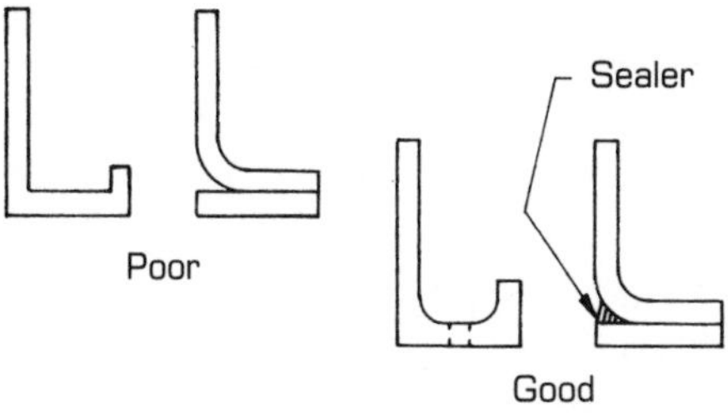

Sealer

Poor

Good

Moisture-collecting pockets should be avoided.

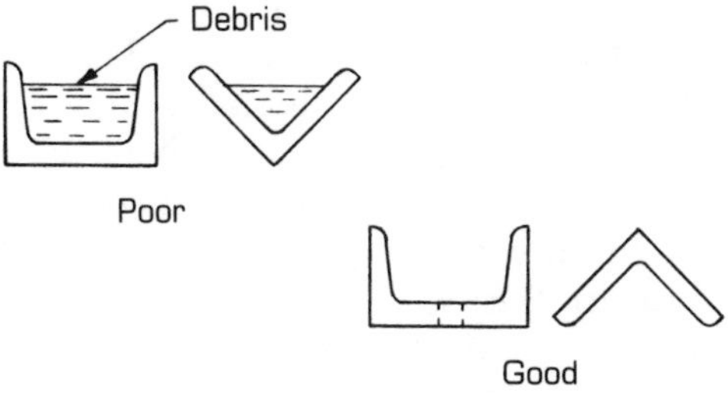

Debris

Poor

Good

Structural members should be arranged so entrapment of debris will not occur.

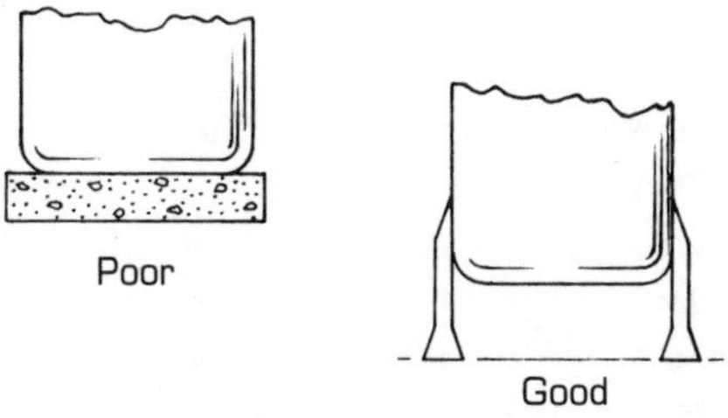

Poor

Good

Storage tanks and other containers should be supported on legs to allow a free circulation underneath. This prevents the possibility of any condensation and collection of moisture under the tank.

TABLE 6-1. Comparison of the Corrosion Rates in a 1% NaCl Solution When Iron Is Coupled to a Second Metal (From Ref 10)

Second metal	Weight loss (iron), mg	Weight loss (second metal), mg	Second metal	Weight loss (iron), mg	Weight loss (second metal), mg
Magnesium....	0.0	3104.3	Tungsten	176.0	5.2
Zinc	0.4	688.0	Lead	183.2	3.6
Cadmium......	0.4	307.9	Tin	171.1	2.5
Aluminum	9.8	105.9	Nickel.........	181.1	0.2
Antimony	153.1	13.8	Copper........	183.1	0.0

Galvanic Corrosion

One of the more serious types of corrosion based upon design occurs when two or more dissimilar metals are electrically coupled and placed in an electrolyte. The ensuing action, known as galvanic corrosion, results from the existence of a potential difference between the metals, causing a flow of current between them. The more active metal undergoes accelerated corrosion whereas corrosion in the less active member of the couple is retarded or eliminated. Bauer and Vogel[10] have shown the change in corrosion rate that results when iron is coupled to each of a series of second metals (Table 6-1).

In the design of a structure involving dissimilar metals, it is essential to know which metal in the couple will suffer accelerated corrosion. The basis for establishing the reactivity of various metals in the couple is the relative electrochemical potential difference between the two. The electrochemical potential for a number of structural metals and alloys is tabulated in the galvanic series shown in Table 6-2. This lists the relative standing of these metals and alloys based upon their potential in a specific environment, in this case seawater.

It has also been shown that the area of the cathodic member of the couple relative to the area of the anodic member is extremely important. The rate of galvanic corrosion is directly proportional to the cathode/anode ratio. For any given current density, the total current is a function of the area. Therefore, when a large cathode is electrically coupled to a small anode, disastrous consequences can result.

A certain anomaly results from this area factor: under certain circumstances, the more corrosion-resistant (cathodic) member of the

TABLE 6-2. Galvanic Series of Metals and Alloys in Seawater

Active	Magnesium
	Zinc
↑	Alclad 3S
	Aluminum 3S
	Aluminum 61S
	Aluminum 52
	Low carbon steel
	Alloy carbon steel
	Cast iron
	Type 410 (active)
	Type 430 (active)
	Type 340 (active)
	Type 316 (active)
	Ni-Resist (corrosion-resisting, nickel cast iron)
	Muntz metal
	Yellow brass
	Admiralty brass
	Aluminum brass
	Red brass
	Copper
	Aluminum bronze
	Composition G bronze
	90/10 Copper-nickel
	70/30 Copper-nickel – low iron
	70/30 Copper-nickel – high iron
	Nickel
	Inconel, nickel-chromium alloy 600
	Silver
	Type 410 (passive)
	Type 430 (passive)
	Type 304 (passive)
	Type 316 (passive)
	Monel, nickel-copper alloy 400
	Hastelloy, alloy C
	Titanium
↓	Graphite
	Gold
Noble	Platinum

couple may require coating in order to minimize the available area and thereby minimize the galvanic corrosion. Fontana and Greene[11] cite the example of a tank that, for economic reasons, was manufactured with carbon steel sidewalls welded to 18-8 stainless steel bottoms. To minimize product contamination, the steel sidewalls were coated with a phenolic resin. Since coatings invariably have pin-hole defects, this resulted in a situation combining a large cathode area (18-8 tank bottom) and a small anode (uncoated regions in the tank walls). The end result was perforation of the sidewalls above the weld.

Another factor worth discussing is that changes can occur in some systems to reverse the polarity of the couple. A good example of this is the system in which aluminum products are fastened with steel rivets. A glance at Table 6-2 shows that steel is more noble than aluminum; therefore, the aluminum should be sacrificed and the steel fastener protected. However, with the onset of corrosion, the aluminum becomes anodized by virtue of the development of an aluminum oxide surface layer, and the polarity reverses. We now have a situation in which the steel fasteners will corrode readily, since not only are they more active than the aluminum structural members but they also exhibit an unfavorable cathode/anode ratio — obviously an undesirable situation! Yet many aluminum stepladders are constructed in exactly that manner today. Figure 6-14(a) shows a close-up view illustrating the use of a zinc-plated tubular steel rivet to fasten the cross braces of an aluminum stepladder to the rear side rails. Obviously a failure here would seriously impair the load-carrying ability of the ladder, yet not only has a galvanic couple been created but the cross section has been significantly reduced by use of a tubular rivet as shown in Fig. 6-14(b). This kind of thinking in design can only result in problems for all concerned — the manufacturer, the seller, and the consumer.

To summarize, in order to minimize attack resulting from galvanic corrosion, the design should wherever possible specify alloys which are compatible with one another based upon a galvanic series. Where this is not possible, every attempt should be made to electrically isolate the members of the couple from one another by the use of gaskets or other methods. Another means which can be used to minimize galvanic attack is to avoid unfavorable area effects resulting from the use of a large cathode and small anode. In a conductive solution, the distance between the dissimilar metals should be made as large as possible. Similarly, the anode should be placed upstream from the cathode to avoid being contaminated by cathodic products being carried downstream. Sealing the bimetallic joint from moisture may be used if other means are not available, but it must be effective or the situation may be compounded by the presence of entrapped corrosion products. Several of these techniques are illustrated in Fig. 6-15.

FIG. 6-14. Steel rivet used to fasten cross brace of an aluminum ladder.

(a) Close-up of riveted area

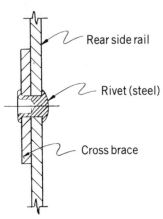

(b) Diagram illustrating cross section

FIG. 6-15. Design considerations useful in minimizing galvanic corrosion.

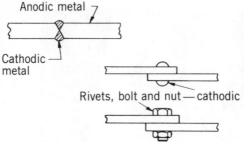

To minimize corrosion attack in butt welded and lap joints, the joint material (weld, rivet, or bolt) should be less active than the larger area metals being joined.

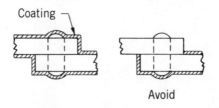

Coatings should be applied with caution. Do not paint the less-noble material; greatly accelerated corrosion will occur at imperfections in the coating. When possible, all exposed surfaces should be painted.

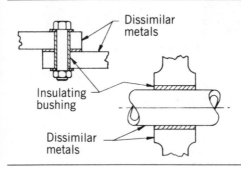

If possible, the connection of dissimilar metals should be separated by an insulating material to reduce or prevent the current flow in the galvanic circuit.

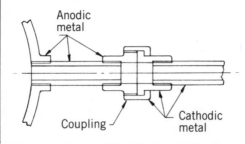

At connections of dissimilar materials, consider using small replaceable sections made of less-noble metal. These expendable parts should be easy to replace, and be made oversize to increase their corrosive life.

Stress Corrosion

Stress corrosion is a mode of failure in which the component undergoes fracture in a delayed and progressive manner. Cracks initiate where tensile stresses are highest and propagate progressively until discovered or until fracture occurs. Stress corrosion as a mechanism has several basic characteristics:

There must be the simultaneous action of stress and corrosion. Tensile stress at the corroding surface is essential and may result from applied stresses in service or residual stresses resulting from manufacturing operations.

There must be a susceptible alloy. Not all alloys are susceptible to stress corrosion cracking. Only certain alloys are, and these are widely cited. Therefore, a review of the corrosion literature is imperative to determine susceptibility.

There must be a specific environment. Even susceptible alloys are specific as to the environment. Not all environments cause problems. Therefore, these sensitive alloys may be used in numerous applications provided that proper precautions are taken.

Because of the complexity of the stress corrosion problem, it is essential as a minimum effort regarding the choice of material that a literature search be conducted in the design stage, relative to the selected alloy and the various environments to which it may be subjected. If the literature search discloses that there may be a potential problem, then another alloy should be used, or the environment modified. Should this not be possible, then a test program should be conducted to determine the stress level at which cracking occurs, the rate of crack propagation, temperature and environmental effects, etc. The nature of the tests to be conducted is in itself the subject of considerable study, largely because there is no single test which fulfills all requirements. Hence, careful consideration must be given to the purpose of the tests, i.e., whether they are designed to perform as screening tests or whether they are intended to serve as predictive tests to determine the resistance of a particular product or design to stress corrosion cracking. One of the more complete references in the area of stress corrosion testing is the work by Craig et al.[12] which deals with the numerous variables in devising a stress corrosion test.

A typical scenario for a stress corrosion failure is as follows. Cracks originate in regions of high tensile stress resulting from either applied loads or residual stresses. This cracking may be preceded by localized attacks such as pitting or intergranular attack which roughen the surface, raising the effective stress and facilitating crack development. At some point, a single crack usually takes over and becomes dominant. With the formation of a crack and the presence of tensile stresses, fracture mechanics theory can be used to describe crack growth. For any single specimen, as the crack length increases by stress corrosion the stress intensity at the crack tip (K) also increases until a critical value K_{IC} is reached, whereupon final fracture occurs by brittle fracture. This behavior is illustrated in Fig. 6-16.

Appearance of Stress Corrosion Failures

Failures occurring as the result of stress corrosion display little ductility and in that sense resemble brittle fractures in macroscopic appearance. There may be several cracks originating from the surface, but failure usually results from the growth of a single crack on a plane perpendicular to the main tensile stress. This type of cracking occurs in many alloys and environments. Table 6-3, while not comprehensive, lists many of the systems where problems have occurred. It is interesting that even though the susceptibility of some of these alloys has been known for as long as fifty years, stress corrosion field failures are still being observed. As an example, let us examine the case of brass flexible gas connectors used in connecting gas appliances to service lines. These connectors are fabricated from thin-walled corrugated brass tubing. The normal installation involves bending

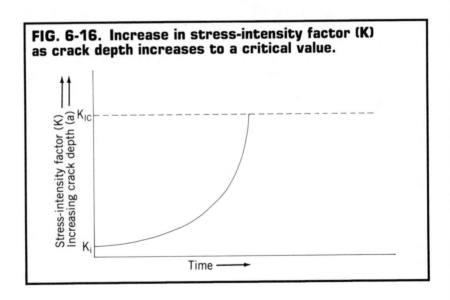

FIG. 6-16. Increase in stress-intensity factor (K) as crack depth increases to a critical value.

TABLE 6-3. Environments That Produce Stress Corrosion in Various Metals (Ref 11)

Material	Environment
Aluminum alloys	$NaCl$-H_2O_2 solutions
	NaCl solutions
	Seawater
	Air, water vapor
Copper alloys	Ammonia vapors and solutions
	Amines
	Water, water vapor
Gold alloys	$FeCl_3$ solutions
	Acetic acid-salt solutions
Inconel	Caustic soda solutions
Lead	Lead acetate solutions
Magnesium alloys	$NaCl$-K_2CrO_4 solutions
	Rural and coastal atmospheres
	Distilled water
Monel	Fused caustic soda
	Hydrofluoric acid
	Hydrofluosilicic acid
Nickel	Fused caustic soda
Ordinary steels	NaOH solutions
	$NaOH$-Na_2SiO_2 solutions
	Calcium, ammonium, and sodium nitrate solutions
	Mixed acids (H_2SO_4-HNO_3)
	HCN solutions
	Acidic H_2S solutions
	Seawater
	Molten Na-Pb alloys
Stainless steels	Acid chloride solutions such as $MgCl_2$ and $BaCl_2$
	$NaCl$-H_2O_2 solutions
	Seawater
	H_2S
	$NaOH$-H_2S solutions
	Condensing steam from chloride waters
Titanium alloys	Red fuming nitric acid, seawater, N_2O_4, methanol-HCl

the connector, with concomitant introduction of tensile stresses into the system.

Some connectors, particularly those used on kitchen appliances such as gas ranges and ovens, are exposed to ammonia-based cleaning compounds, thereby creating a very favorable stress corrosion condition. Perforation of the connector through a stress corrosion crack would cause gas leakage with the potential for fire and explosion. A

FIG. 6-17. Stress corrosion crack (arrow) in a flexible gas connector.

connector which has suffered just such a stress corrosion crack is shown in Fig. 6-17. This type of failure can be avoided by coating the external surfaces of flexible connectors with protective polymers. Though the current generation of connectors is manufactured with this precaution, countless numbers of uncoated connectors are still in service and present an enormous hazard.

HYDROGEN DAMAGE

Hydrogen damage is a phenomenon which has been encountered and recognized for many years. It is most commonly associated with metals and alloys having a body-centered-cubic (BCC) crystal lattice structure, particularly high strength steels. Hydrogen can also produce degradation in other metal systems, such as titanium and zirconium alloys and some refractory metals, e.g., tungsten, vanadium, tantalum, and columbium. While the results of hydrogen damage are no less devastating than those of stress corrosion, the economic consequences are not as serious because of the lower utilization. Obviously, the susceptibility of these high strength steels represents a serious vulnerability which must be considered during design and subsequent manufacturing processes.

Hydrogen degradation may manifest itself in several forms, e.g., internal flakes or bursts, cracks, blisters, or decreased mechanical properties. Hydrogen-affected metals generally exhibit a loss of tensile strength and ductility which becomes more pronounced as the testing temperature decreases or a strain rate increases. Poorer impact strength and toughness can also result from increased hydrogen concentrations. Increasing the strength level of the material causes increased susceptibility as indicated by a decrease in the time to failure at any particular stress level.

Additionally, in material with high hydrogen concentrations, the applied or residual stress required to cause failure decreases as the stress intensity of the notch increases, a manifestation with serious implications in structures containing geometrical or fabrication-related stress raisers. Troiano[13] has proposed that under the influence of an applied stress hydrogen would diffuse toward regions in a triaxial stress state. Furthermore, these regions of triaxiality would result from a network of internal voids acting as internal stress raisers. The hydrogen which exists in the stressed region of the lattice near the voids may donate its electron to the metal, thereby decreasing the binding energy or cohesion of the lattice in this region. Hydrogen uniformly distributed throughout a metal lattice is nondamaging because its concentration is so small. But when hydrogen is segregated in highly stressed regions by stress-induced diffusion, the result is a lowering of the cohesive strength of the metal locally, with subsequent crack propagation and embrittlement.

In a study of the effects of high pressure hydrogen on the mechanical properties of 304 stainless steel, Vennett and Ansell observed that the degradation noted was directly related to martensite transformation resulting from plastic deformation.[14] As a result, they concluded that lattice hydrogen, rather than adsorbed hydrogen, was by far the most damaging. They further suggested that hydrogen damage might result from a combination of mechanisms rather than from a single specific mechanism, depending on the material and the circumstances.

Appearance of Hydrogen Damage

As previously stated, hydrogen damage may appear as internal flakes, cracks, blisters, or decreased mechanical properties. Flakes are internal fissures which may occur on the interior of a forging on a plane parallel to the forging direction. A forged billet exhibiting several large hydrogen cracks is shown in Fig. 6-18. These cracks may be detected by ultrasonic inspection, by dye-penetrant inspection of the billet ends, or by etching transverse sections as shown in Fig. 6-19.

FIG. 6-18. Forged billet exhibiting hydrogen cracks, dye penetrant tested (3×).

Typically, flaking is a problem in large air-melted forgings of high alloy steel such as turbine or ship shafts. As a consequence, these products are now usually produced by melting techniques which reduce the hydrogen level, such as vacuum degassing and vacuum arc remelting.

Also, as previously mentioned, one of the principal effects of hydrogen embrittlement in steels is a decrease in tensile strength and ductility when tested under static loads or low strain rates. According to Tetelman,[15] impact tests do not indicate susceptibility and, for this reason, are not recommended to determine whether embrittlement exists. This view is not universally held, however, and there have been some instances in which poor impact properties were associated with high hydrogen concentrations.

On a fractographic basis, the observed features vary. Micro-cracks initiate internally, often near inclusions or other interfaces, and propagate intergranularly for an indeterminate distance. They may also originate from electroplated surfaces and propagate into the interior of the component. In high strength steels, the

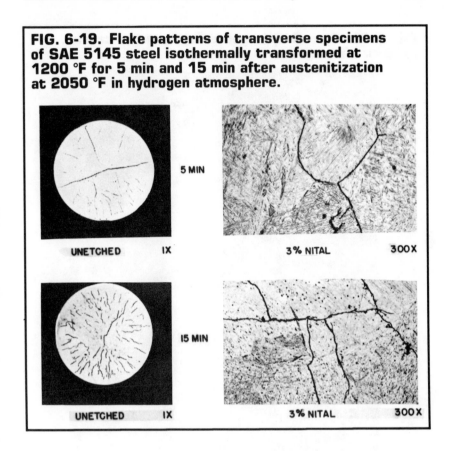

FIG. 6-19. Flake patterns of transverse specimens of SAE 5145 steel isothermally transformed at 1200 °F for 5 min and 15 min after austenitization at 2050 °F in hydrogen atmosphere.

hydrogen-caused cracks are intergranular in nature when examined either metallographically or with the electron microscope. A typical hydrogen crack is shown in Fig. 6-20. Additional examples can be found in various reference books dealing with failure analysis.[16,17]

Sources of Hydrogen

The origin of hydrogen in the final product is dependent on the nature of the product. In large forgings, the source may be traced back to the original melting. A major source of hydrogen in steels can be the water vapor present in the atmosphere as a result of high humidity from damp scrap metal used to charge the furnace or from moisture-contaminated slag. This vapor reacts with the molten steel at these elevated temperatures producing hydrogen which goes into solution.

FIG. 6-20. Typical hydrogen crack, in 4140 steel (500×). 2% nital etch.

As the steel solidifies after pouring, the solubility decreases and the resultant hydrogen becomes trapped.

From the standpoint of product liability this kind of problem, which primarily affects forgings, can be traced to the original steel manufacturer. However, intermediate processors have an obligation to prevent these forgings from entering the marketplace through the use of nondestructive testing.

The most common source of hydrogen is that occurring as the result of electroplating or electrolytic cleaning operations as described in Chapter 6. Embrittlement as a result of this kind of exposure usually occurs late in the manufacturing cycle and may not be

FIG. 6-21. Section from a failed accumulator ring (1.5×). Note fine radial cracks (arrow). See also Fig. 6-22.

discovered until after the product is shipped as illustrated by the following example. Fracture of an accumulator ring forged from 4140 steel was discovered during inspection and disassembly of a hydraulic-accumulator system stored at a depot. The ring had broken into five small and two large segments.

The small segments of the broken ring displayed very flat fracture surfaces with no apparent yielding, but the two large segments did show evidence of bending (yielding) near the fractures. In addition, some segments contained fine radial cracks (see arrow on segment shown in Fig. 6-21). Inspection of fracture-surface replicas by electron microscopy disclosed the fracture mode to be predominantly inter-granular (Fig. 6-22). Also observed were many hairline indications on the fracture facets and some partly formed dimples. All these features are indicative of the cracking associated with hydrogen embrittle-ment. Gas analysis conducted on the specimen yielded hydrogen values ranging from 0.15 to 2.29 ppm, indicating a localized hy-drogen problem. Review of the processing revealed that the compo-nents required cadmium plating, after which they were scheduled for stress relieving at 260 °C (500 °F) for three hours. An investigation showed that there was strong evidence that one batch of rings did not

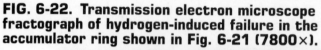

FIG. 6-22. Transmission electron microscope fractograph of hydrogen-induced failure in the accumulator ring shown in Fig. 6-21 (7800×).

receive the required stress-relieving treatment. Had these defective rings not been caught while in storage, they could have entered the distribution system with the potential of causing serious damage in service.

Exposure in service can also cause hydrogen embrittlement. Process fluids bearing hydrogen or sour petroleum oil can cause embrittlement and premature failure of the pumping and handling equipment.

Welds can also be affected by hydrogen. Underbead cracking occurs readily when high concentrations of hydrogen are present. Moisture is the usual source of hydrogen, in this case being present in the atmosphere as the result of high humidity, in the electrode coating or on the metal surfaces. Procedures to eliminate moisture, such as preheating, baking of electrodes, and gas shielding, are essential when welding high strength steels.

WEAR

Wear can be defined as the deterioration of a surface as a result of use, the ordinary condition of use being the movement of one surface in contact with another. Wear is affected by a variety of factors such as the type and mode of loading, speed of surface movement, type and degree of lubrication, surface treatments, temperature, surface finish, hardness of materials in contact, presence of contaminants, and the chemical nature of the environment. All these factors can and do affect the rate of wear and therefore should at least be considered when a design is expected to be in a wear situation. While a total consideration of all the factors affecting wear is beyond the scope of this book, we shall attempt to summarize the general effects which can be expected.

Lubrication

The purpose of a lubricant is to minimize the contact between the sliding surfaces, thereby reducing the coefficient of friction. In terms of service conditions, one must distinguish between boundary lubrication and hydrodynamic lubrication. Figure 6-23 illustrates the two regions for the case of simple bearings. The coefficient of friction and film thickness are plotted against the parameter ZN/P, where Z is viscosity, N is rpm, and P is load. To the right of the dashed vertical line is the region of hydrodynamic lubrication, where the surface asperities are completely separated by an oil film of such thickness that no metal-to-metal contact can occur. In this region, therefore, the properties of the lubricant are of primary importance. To the left of this line is the region of boundary or thin-film lubrication. As seen in Fig. 6-23, the film thickness in boundary lubrication is so small that asperities can, and do, come into contact through the oil film. In this region of true metal-to-metal contact the properties of the metals are of primary importance.

Solid lubricants are particularly valuable for applications where low velocity, high pressure, or high temperature renders liquid lubricants ineffective. Most solid lubricants such as graphite and molybde-

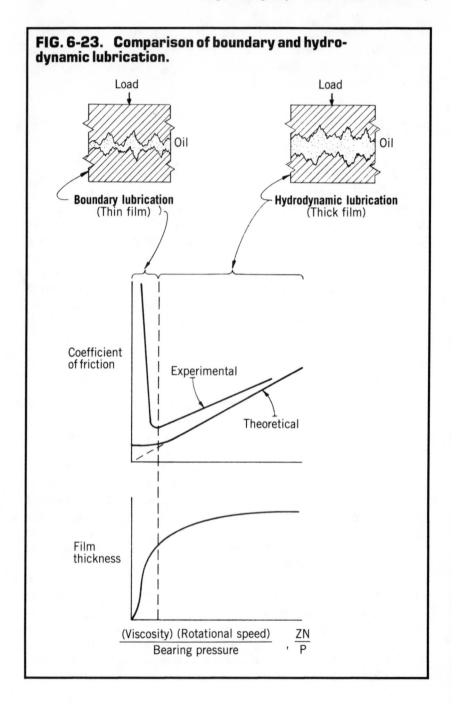

FIG. 6-23. Comparison of boundary and hydro-dynamic lubrication.

num disulfide have a layer-lattice crystal structure with layers of hexagonal platelets that are solidly linked in the basal plane but allow slip between layers. A solid lubricant should possess two characteristics: a low shear strength to give a low coefficient of friction and good adherence to the surface. Wear, not friction, is usually the main problem in design. A lubricant that forms a continuous adherent layer and thus prevents metal-to-metal contact will give low wear.

Surface Treatments

In addition to lubrication, numerous surface treatments are utilized in an attempt to minimize the effects of wear. Table 6-4 summarizes the types of methods in use and their suitability for various wear mechanisms.

Effects of Temperature

Increasing the service temperature increases the wear rate largely as the result of two factors. First, there is a decrease in the hardness of the contacting surfaces with a corresponding decrease in the yield strength. This tends to increase the probability of welding and plastic deformation at the surface asperities while they are in contact. Second, where there is a corrosive aspect to wear, the corrosion rate increases with increasing temperature. Obviously, then, it is a design error if you design based upon a wear rate at one temperature range and the component in service experiences a higher temperature.

Surface Finish

As a general rule, the higher the surface roughness the higher the resultant wear rate. However, extremely smooth surfaces tend to suffer more from the presence of abrasive contaminants, since these can readily attack the surface and cannot be lost and hidden in surface irregularities.

Influence of Hardness

As the hardness of the relative surfaces increases, the wear rate generally decreases. The hardness should be the result of improvements resulting from alloying or heat treatment, since experience has shown that work hardening the material raises the hardness but does little to improve the wear life. Rabinowicz[18] states that there are four

TABLE 6-4. Factors in Selection of Surface Treatments for Resistance to Various Wear Mechanisms

Wear mechanism	Surface requirement	Surface treatment	Conditions influencing selection
Abrasive wear	High hardness, adequate toughness, moderately thick coating	Weld deposits, thermal and thermochemical treatments	**Abrasive or abrading hardness:** Coating selected must be harder than abrasive or contacting surface hardness but little advantage in increasing hardness above 1.5 times abrasive hardness. **Size of abrasive particles:** Tougher materials required for larger-sized abrasives. Harder materials for smaller and/or harder abrasives.
Adhesive wear	High hardness or dissimilar composition to contacting surface, or be nonmetallic	Thermo-sprayed alloys, electroplated Cr, thermochemical treatments, plasma-sprayed ceramics and carbide/cermets, electroless Ni, TiC (by CVD)	**Degree of lubrication:** Soft dissimilar metal coatings can be used where lubricant is present. Harder materials necessary when lubrication is marginal or absent. Hardness requirement determined by contact pressure and hardness of contacting surface. **Surface roughness:** Wear greater with rougher surfaces. Nonmetallic surfaces, produced by phosphating and nitrocarburizing, prevent wear during "running in," at high loads and speeds.

Contact fatigue	High yield strength, adequate toughness, thick coating, good bond strength	Thermal and thermochemical hardening, weld deposits, sprayed and fused coatings	**Applied stress:** Nitrided case or thick case, usually obtained by carburizing but sometimes by nitriding, required at high operating stresses. At low stresses the cases obtained by nitrocarburizing are adequate.
Erosion	High hardness, adequate toughness, good bond strength	Weld deposits, plasma-sprayed carbide/cermets and metals	**Velocity of impact:** At low velocities (300 ft/sec) elastic materials (rubbers) can be used but at medium and high velocities (800 ft/sec) combination of hardness and toughness required.
Fretting	High hardness, or dissimilar composition to contacting surface, or be nonmetallic	Sprayed Cu alloys; anodizing (on Al and Ti), nitrocarburizing (on steels), plasma-sprayed oxides and carbide/cermets.	**Corrosion resistance:** Material should be corrosion resistant. Use dissimilar materials whenever possible — i.e., when galvanic corrosion is not a problem.
		Dry film lubricants	Increase load to stop motion, but do not overload.

primary wear mechanisms: adhesive wear, abrasive wear, pitting and spalling, and corrosive attack. The first two account for the majority of wear situations, with adhesive wear being predominant. Adhesive wear, also referred to as galling, results from the welding together and subsequent shearing of asperities when surfaces in contact are slid past each other. Abrasive wear occurs when two surfaces, one of which is harder and rougher than the other, are in sliding contact. Similar damage can result when hard abrasive particles are embedded in a softer matrix. The damage occurs because of the plowing action of the hard particles or asperities, a quasi-machining process that results in a substantial loss in surface material.

The primary mechanism of wear in metals and alloys at high temperatures is the development of localized hot spots at the points of contact with plastic flow associated with the contact regions. When two different materials are in sliding contact, the material with the lowest yield point at temperature will show the greatest amount of wear.

Contamination

With few exceptions, contaminants increase the rate of wear. Contaminants which commonly cause problems are abrasive particles (sand, for example), moisture, and chlorides.

REFERENCES

1. Orowan, E., *Trans. Inst. Eng. Shipbuild.*, Scotland, 1945, Vol. 89, p. 165.
2. Irwin, G. R., *Encyclopedia of Physics*, Vol. 6, Springer, Heidelberg, 1958.
3. *Fracture Prevention and Control*, American Society for Metals, 1974.
4. Memorandum, Sept. 13, 1966, from V. J. Burns, Deputy Chief Engineer, N.Y.S. Dept. of Public Works, to J. A. Hanson, Division Engineer, U.S. Bureau of Public Roads, Albany, N.Y.
5. Buxbaum, O., Service Fatigue Loads Monitoring, Simulation and Analysis, STP 671, American Society for Testing and Materials, Philadelphia, 1979, pp. 5-20.
6. Abelkis, P. R., "Fatigue Loads," *ASTM Standardization News*, Vol. 8, No. 2, Feb. 1980, p. 19.
7. P. G. Fluck, *Proc. ASTM*, Vol. 51, 1951, pp. 584-592.
8. "What You Should Know About Metal Fatigue," *Iron Age*, Nov. 26, 1979, pp. 131-136.
9. "VAR Materials Are Back in the Saddle Again," *Iron Age*, Oct. 1, 1979.
10. O. Bauer and O. Vogel, *Mitt. Deut. Materialprüfunganst.*, Vol. 6, 1918, p. 114.

11. M. G. Fontana and N. D. Greene, *Corrosion Engineering*, McGraw-Hill, New York, 1967, p. 2.
12. Craig, H. L., Sprowls, D. O., and Piper, D. E., *Handbook on Corrosion Testing and Evaluation*, John Wiley, New York, Chap. 10, "Stress Corrosion Cracking," pp. 231-290.
13. Troiano, A. R., Campbell Memorial Lecture, *Trans. ASM*, Vol. 52, 1960, p. 54.
14. Vennett, R. M., and Ansell, G. S., *Trans. ASM*, Vol. 60, No. 2, June 1967, pp. 242-251.
15. Tetelman, A. S., and McEvily, A. J., Jr., *Fracture of Structural Materials*, John Wiley, New York, 1967, p. 456.
16. *Metals Handbook*, 8th Ed., Vol. 10, American Society for Metals, Metals Park, Ohio 44073, 1975, pp. 230-240.
17. Colangelo, V. J., and Heiser, F. A., *Analysis of Metallurgical Failures*, John Wiley, New York, 1974, pp. 224-238.
18. Rabinowicz, E., *Friction and Wear of Materials*, John Wiley, New York, 1965, p. 113.

7

Primary Manufacturing Processes That Affect Product Liability

INTRODUCTION

Following a successful design period, the various stages of which have been elaborated in Chapters 4 and 5, fabrication or construction of the product generally begins. Initially, the component may start as a casting or a pre-formed product such as forgings, rolled bar and plate, extrusions, drawn wire, powder metallurgy compacts, stampings, or a variety of nonmetallic materials such as composites, polymers, ceramics, etc., which may require more elaborate initial fabrication. Regardless of its origin, the starting material will very likely undergo further processing which will change its shape and corresponding properties. Such processing may be relatively simple — sawing, grinding, or perhaps just painting. However, it is the components that are complex in geometry and material that we are mainly interested in, because they usually create processing problems and require extensive or involved treatments, circumstances that may seriously affect the product liability position of a company.

Since virtually all manufacturing processes change either the shape or the physical characteristics of an engineering material, they must be performed with proper consideration for the changes that are incurred. Certain manufacturing processes can be beneficial to the properties of a material, improving its performance and reliability. Conversely, some processes can be deleterious to a material, particularly if they are not properly controlled. Furthermore, what's good for one material may not necessarily benefit another, so material processing sequences cannot be universally applied in manufacturing. Defective conditions produced in a material or a component during manufacturing processing can create a hazardous or unreliable product, even though it meets the drawing and design requirements. Often such hazards or defects are detected during subsequent inspections. Some manufacturing operations include a quality control examination after each major processing stage. This procedure can isolate a problem and allow corrective action to be initiated early in the manufacturing operation. Early detection of a problem also avoids the expenditure of further labor, energy, and materials on components that are not usable or salvageable.

The Legal Aspects

In this chapter and Chapter 8 we will be dealing with flaws or defective conditions caused by the processing operations during manufacture of parts and materials. The question that frequently arises in connection with manufacturing defects can be simply stated: Does the defect or flaw constitute a product liability problem? According to Weinstein et al.,[1] if a plaintiff sues in a product liability case he must establish the following criteria:

1. The product was defective.
2. The defect existed at the time the product left the defendant's hands.
3. The defect actually caused the injury.
4. The injury was due to the identified defect (not to some other defective condition that may co-exist in the product).

It follows that the processor or manufacturer should be aware of *what constitutes a defect* in regards to product liability cases. We will not attempt to define a defect in legal terms, but rather point out the procedures used by the legal and judicial fields to define defects. As Weinstein et al. point out, the critical question is "whether as a result of the flaw, the product performs in a *substandard* manner, and if so, does such performance make the product defective?"

Weinstein et al.[2] further report that in cases involving a production-related defect, it is not difficult to identify the defect — obviously not the most heartening news for manufacturers or processors. During the investigation a comparison is usually made between the defective product and an acceptable or good product from the defendant's own assembly line. The reference standard, then, is the manufacturer's own internal quality standard. If the product in question fails to measure up to that standard, it can, the authors assure us, be identified as defective.

In light of this analysis, it is possible to extend this comparison to the same or similar components produced by other manufacturers and processors. Furthermore, the part in question may be reviewed against any industry-wide standards and codes that may be applicable. Once the part has been identified as defective and found responsible for causing the plaintiff's injury, a *prima facie* case has been established. The key point we wish to emphasize is that the quality standards set by a manufacturer or processor may well be used to assess the presence of a defect in his own product!

Although it is important to keep accurate and thorough documentation with regard to quality control, such as reports on nonconforming material, rejection reports, and waivers for deviation as well as acceptance records, remember that these documents may be subject to subpoena for a product liability litigation. Under certain circumstances, these records could be very incriminating. For example, consider the hypothetical situation of a component that was put on waiver by the inspection personnel for insufficient material properties. Perhaps this insufficiency was slight and could be rationalized or explained without considerable effort, or maybe, because the part was sorely needed to fill an important order, it was eventually accepted. Then this particular part was involved in a failure which resulted in serious property damage or injuries. Even if the part was not to blame, or was not directly involved in precipitating the failure, the records will show that it was substandard. Most assuredly, this is not a strong position to argue from.

Another adverse situation can possibly arise when a component involved in a product liability suit exhibits characteristics which the quality control records show have been the basis of prior rejections or reports on nonconforming materials. Generally, an examination of the records will disclose this fact without too much effort. Terms like "unacceptable," "rejected," "nonconforming," etc., should not be used to describe components or products which may be acceptable after reworking or further processing. Such terminology in the official records (manufacturing and quality) carries high potential for product liability.

If the foregoing discussion sounds like an argument for record-keeping subterfuge, it is not. What we are really suggesting is that accurate and thorough records be kept on components that do not meet acceptance standards the first time out. Many nonconforming parts may eventually be pronounced acceptable through reinspection, retesting, and reprocessing or repair. Although such techniques are customary in most industries, the manner in which the final acceptance is judged — in other words, the criteria that were used — must be explicitly documented. Such documentation enables a producer to defend his decisions for marketing a part, if that becomes necessary in a product liability case.

Objectives

The object of this chapter, therefore, is to emphasize and discuss just such circumstances: the processing problems that create defective or hazardous materials and components and increase the product liability potential of a manufacturer. Armed with the knowledge that a problem can occur in a particular process, the manufacturing and engineering personnel can take steps to avoid or at least minimize it. Accurate documentation and record keeping will help to minimize the product liability exposure of a company in these matters as well as provide the engineering and quality control departments with necessary information if a problem does arise.

We have encountered numerous manufacturing problems that could have been resolved much more expediently if the necessary processing data had been available before the product defect became apparent. The resolution of a production problem is seriously impeded if one has to begin to generate fundamental processing data after a manufacturing problem has occurred.

As emphasized in previous chapters, the threat of product liability decreases for a company when defects and hazardous conditions are alleviated. Just how significantly it decreases depends largely on the individual circumstances of a product or a situation. The design stage of production is very important with respect to product liability. We will now demonstrate that the processing stages of manufacture are also very important in avoiding future product liability problems.

MELTING AND SOLIDIFICATION

The initial process stage of any engineering alloy is melting and casting, regardless of whether the material is poured intrinsically as a casting or as an ingot that will subsequently be mechanically worked into another shape. Many factors influence the melting process and fortunately most can be adequately controlled. However, problems may arise at this early stage which can eventually result in defects or defective material conditions in the final product. For example, the cleanliness or purity of the initial furnace charge is very important in the production of high quality alloys. It is much less of a problem to control additions to a melt than it is to remove contaminants and tramp elements that were incorporated in the scrap or melting stock constituting the charge. Also, second phase particles can be picked up by the melt from the refractory linings on the furnace and transfer vessels. These materials have a higher melting temperature than the alloy and thus are not readily dissolved. Such *exogenous* inclusions occasionally end up in the final product and, depending on their nature and severity, have been associated with adversely affecting the following properties: formability, fatigue life, fracture toughness, machinability, mechanical properties in general (particularly trans-verse ductility and toughness in forged or rolled products), and resistance to quench cracking during heat treatment. Nonmetallic inclusions with their attendant ramifications can sometimes be re-garded as a *latent defect* — in other words, a deleterious condition that was not obvious or easily detectable during previous inspections but is discovered at a later stage of processing or in service.

An example of exogenous nonmetallic inclusions in low alloy steel is shown in Fig. 7-1. These inclusions were directly responsible for the failure of a large pressure vessel during a manufacturing operation which dilated the container. Electron beam analyses and X-ray methods identified the constituents of these inclusions to be zirconium, oxygen, silicon, and manganese. These formations were eventually traced to spalling of the refractory linings of the furnace and ladles in the melting and pouring operation. We will explore the harmful effects of inclusions in much greater detail later in this chapter, under the heading Ingot Defects.

Casting Defects

Once the liquid metal alloy is poured, another set of problems, associated with solidification or freezing, can occur in the material. Some of these flaws, typically known as casting defects, include gas porosity, shrinkage porosity or cavities, cold shuts, scabs, sand holes, etc. These imperfections have been defined and thoroughly discussed in the technical literature.[3-5] Therefore, it is sufficient to state that such flaws can usually be attributed to casting design and to foundry practice. In addition, if these flaws go undetected, they may eventual-ly be initiation sites for failures during service. The mechanisms for such failures were presented in Chapter 6. Several examples of cast-ing imperfections are illustrated in Fig. 7-2.

FIG. 7-1. Exogenous nonmetallic inclusions in low alloy steel.

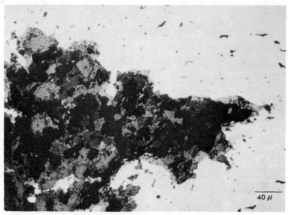

40 μ

(a) Complex inclusion with dispersion of smaller particles in surrounding matrix

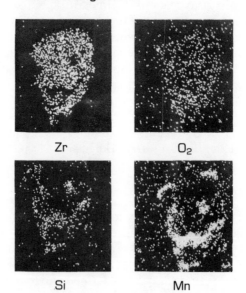

Zr

O₂

Si

Mn

(b) X-ray images of various elements in these inclusions

FIG. 7-2. Examples of casting imperfections along with their possible causes and remedies. (Reproduced by permission, American Technical Society)

DEFECT	PROBABLE CAUSES	REMEDY
1. SHRINKAGE 	1. IMPROPER FEEDING 2. LACK OF PROGRESSIVE SOLIDIFICATION	1. USE ADDITIONAL RISERS OR A COMBINATION OF CHILLS AND RISERS 2. MODIFY GATING TO SUPPLY COOLER METAL TO AREAS WHERE SHRINKAGE IS 3. MODIFY DESIGN FOR MORE UNIFORM WALL THICKNESS
2. POROSITY 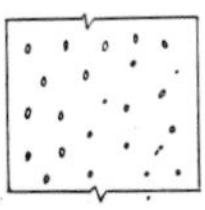	1. DISSOLVED GASES 2. EVOLVED GAS FROM DAMP SAND OR VOLATILE BINDERS 3. MECHANICALLY ENTRAPPED GASES 4. SHRINKAGE MICROPOROSITY	1. USE VACUUM OR FLUX DEGASING 2. USE SAND WITH PROPER PERMEABILITY, PROPER BAKING, AND LOWER MOISTURE 3. CHANGE GATING TO REDUCE MECHANICAL AGITATION 4. REDUCE POURING TEMPERATURE 5. USE CHILLS TO ACCELERATE SOLIDIFICA-TION
3. MISRUNS AND COLDSHUTS 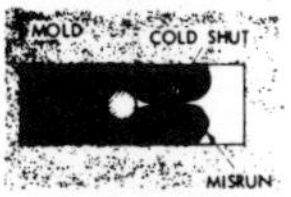	1. LOW FLUIDITY OF MOLTEN METAL	1. INCREASE POURING TEMPERATURE 2. ALTER GATING SYSTEM TO PROVIDE EASIER ENTRY TO MOLTEN METAL INTO MOLD CAVITY 3. CHANGE TO AN ALLOY HAVING GREATER FLUIDITY
4. SEGREGATION, CORING & DENDRITIC GROWTH 	1. POOR AGITATION IN MELTING OR HOLDING FURNACES 2. EXCESSIVELY SLOW COOLING	1. PROVIDE PROPER AGITATION AND MIXING, i.e. INDUCTION HEATING 2. MODIFY GATING AND CHILLING TO IMPROVE SOLIDIFICATION CONDITIONS
5. EXCESSIVE GRAIN SIZE 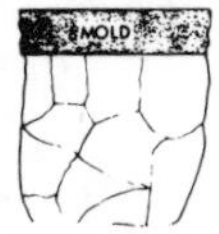	1. EXCESSIVE POURING TEMP. 2. PROLONGED SOLIDIFICATION TIME	1. REDUCE POURING TEMPERATURE 2. USE MOLD MATERIAL WITH HIGHER HEAT CONDUCTIVITY 3. USE INOCULANTS
6. CRACKING (HOT TEARS) 	1. HOT SHORTNESS 2. EXCESSIVE RESTRICTION OF CASTING DURING CONTRACTION	1. MINIMIZE RESISTANCE TO SHRINKAGE BY USE OF CORES HAVING GREATER COLLAPSIBILITY 2. STRENGTHEN SECTIONS SUBJECT TO CRACKING 3. CHANGE CHILLING, GATING, AND RISER-ING TO REDUCE HOT SPOTS
7. INCLUSIONS 	1. REFRACTORIES 2. SAND FROM MOLD OR CORE 3. OXIDE FILMS AND DROSS	1. CLEAN EQUIPMENT 2. HANDLE MOLDS AND CORES CAREFULLY 3. HOLD MELT WITHOUT AGITATION FOR A SHORT TIME PRIOR TO POURING 4. AVOID EXCESSIVE AGITATION OF THE MELT

Usually, external casting defects are easily detected, particularly after the castings are cleaned by sandblasting or rotoblasting. Often these problem areas can be alleviated by machining or grinding and weld repair. However, if they are internal, such as shrinkage cavities, gas pores, or nonmetallic inclusions, they must be nondestructively detected by radiography or ultrasonic inspection. Depending on the size and location of such defects, the casting may be salvaged or scrapped. The key points are *detection* and *identification*. If casting imperfections exist, they should be identified and analyzed for possible contribution to the performance of the final product. These imperfections in themselves may not result in a failure but most certainly can act as the origination sites for crack initiation and crack growth under conditions of fatigue or environmental attack. Recalling our discussion of stress concentrators in Chapter 5, the stress-concentrating effects of these defects depend upon their *shape, size,* and *location* in the casting.

In view of the present nondestructive test methods available for quality control purposes, detection of most consequential defects is possible. For example, the sensitivity of radiography is on the order of 2% of the thickness being analyzed.[6] In other words, a flaw as small as 2% of the section thickness is detectable. Presently, even 1% sensitivity is possible with special radiographic techniques. Therefore, the means to identify and analyze casting flaws are available to foundry and casting producers. As we previously emphasized, the presence of stress concentrators such as casting defects can have a detrimental effect on the integrity and reliability of an engineering component or structure. While these imperfections may be difficult to eliminate, good casting design and sound foundry practice are the first avenues of control. Second, detection and analysis of such flaws will significantly reduce the chances of putting defective products in service, thus minimizing the product liability potential of the casting producer and subsequent processors.

Ingot Defects

In a manner similar to that in castings but usually on a larger scale, solidification of a liquid metal alloy in an ingot mold produces a class of imperfections due to reactions in the liquid and the solid state. Typically being large (on the order of tons), ingots tend to freeze slowly, thereby promoting the redistribution of solute or alloying elements.[7] The final result is a variation in chemical composition throughout the solidified ingot, as illustrated in Fig. 7-3.

FIG. 7-3. Chemical segregation revealed on macro-etched cross section. Ring patterns result from differential etching of segregated regions.

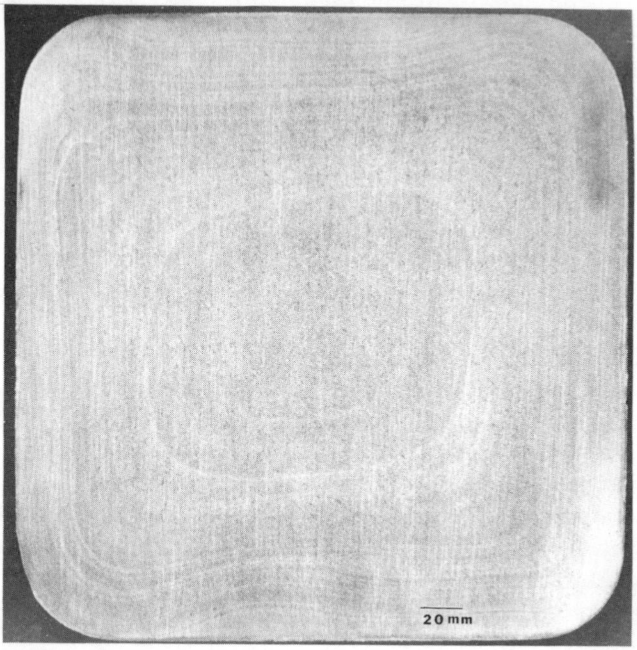

Chemical Segregation. Nonuniform distribution of alloying elements is commonly referred to as *chemical segregation* and can be observed on both a microscopic scale (on the order of dendrite spacings) and a macroscopic level (on the order of centimeters). Segregation has frequently been associated with a variety of materials problems such as poor transverse ductility and toughness in mechanically worked products, large variations in mechanical properties within a component or within lots, heat treatment problems such as incomplete austenitization or solutionizing and quench cracking, and, also, differences in machinability (hard spots).

Chemical segregation in alloys is not only difficult to prevent but also very difficult to alleviate once the condition exists. For instance, very high temperature homogenizing treatments are sometimes successful in reducing segregation but often require furnace times that are impractical and uneconomical. In fact, it has been reported that large-scale carbon segregation in steels is not appreciably affected for days or years at ordinary austenitizing temperatures.[8] Indeed, temperatures approaching the melting point may be necessary to effect adequate diffusion, and the potential for constitutional liquation (localized melting of low temperature phases) becomes significantly greater.

Although ingot size depends heavily on the eventual mechanically worked shape, smaller ingots should at least be considered in the production analysis (process design) because they tend to cool faster, thereby providing less opportunity for chemical segregation. Other alternatives, if warranted, include *consumable remelting* techniques such as electroslag refining (ESR) and vacuum arc remelting (VAR).[9-12] Because both of these melting techniques involve a much smaller liquid pool than statically cast large ingots, segregation is minimized.

It is worth mentioning that both ESR and VAR are refining techniques that usually result in a cleaner material with substantially fewer nonmetallic inclusions than an air-melted product of the same grade. Correspondingly, the mechanical properties of the refined materials are better, particularly impact strength, ductility, and fracture toughness.[13] Furthermore, vacuum treatment of alloys for improvement of properties is not restricted to steels: it is also applicable to nonferrous alloys, including aluminum, copper, and titanium base materials.[14]

Other deleterious conditions that may occur during solidification include *piping, centerline shrinkage, hydrogen flaking,* and *endogenous nonmetallic inclusions,*[15] plus the gas pore formation and shrinkage conditions previously discussed in this chapter under Casting Defects. Let us now examine the four primary conditions listed above.

Ingot Piping and Centerline Shrinkage. This specific type of defect occurs in the top central portion of an ingot and is associated with contraction of the metal during solidification. The resulting cavity is irregularly shaped, but generally appears as an inverted cone. Piping is illustrated in Fig. 7-4. In addition to the primary pipe, secondary regions of pipe and centerline shrinkage may extend deeper into an ingot as illustrated in Fig. 7-5.

Although primary piping is usually an economic concern rather than a material problem, if undetected it can eventually result in a defective forging or rolled product. Piping can be minimized by pouring molds with the big end up and applying sufficient hot-top material (exothermic) immediately following the pour.

FIG. 7-4. Schematic illustration of ingot piping. Ingots poured big end up (BEU) and big end down (BED), without hot topping.

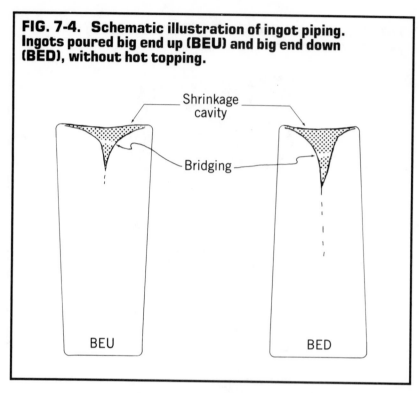

On the other hand, secondary piping and centerline shrinkage can be very troublesome, since they may escape detection in the mill and subsequently cause centerline defects in bar and wrought products. Such a condition may indeed provide the initial stress concentrator for a future fatigue failure in the case of a cyclically loaded component, or a central rupture in a Mannesmann-type loading situation in subsequent processing.[16]

Hydrogen Flaking. Hydrogen damage to metals can take many forms. However, we will be concerned with the segregation of hydrogen to local regions of an ingot. This form of defect is commonly referred to as *flaking* and results in internal cracks or fissures. Microcracking associated with hydrogen flaking in a low alloy steel is shown in Fig. 7-6.

A major source of hydrogen in steels results from the reaction of water vapor with the liquid metal at very high temperatures. The water vapor may originate from the scrap used to charge the furnace, the slag ingredients and additions, the refractory materials lining the

FIG. 7-5. Longitudinal section of ingot showing extensive centerline shrinkage.

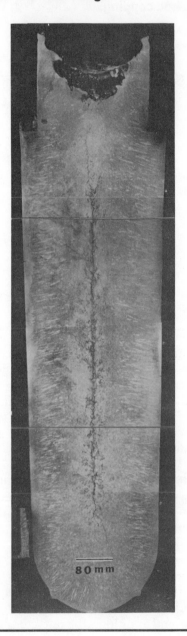

FIG. 7-6. Microcracking associated with hydrogen flaking in low alloy steel: (top) longitudinal, (bottom) transverse.

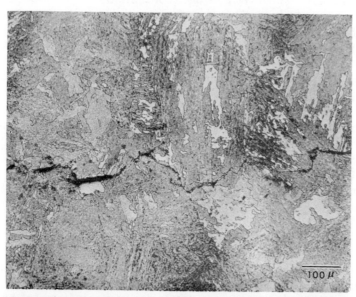

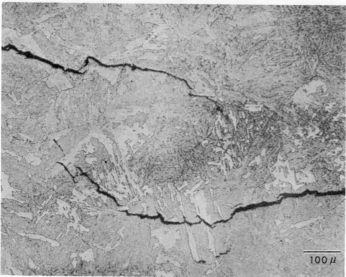

transfer vessels, or even the ingot molds themselves. The resulting hydrogen can be entrapped in the metal lattice during solidification. Hydrogen concentrations in excess of about 5 ppm (0.0005%) have been associated with flaking, especially in heavy sections and higher carbon steels.[17] However, hydrogen concentrations in excess of only 1 ppm (0.0001%) have been associated with degradation in impact toughness in certain steels.

In cases where hydrogen flaking can serve as the initiation site for cracking and thus increase the potential for future problems, it would be prudent to employ a thermal treatment which can relieve this condition. For example, slow cooling after the mechanical working operation or a separate annealing cycle would relieve any internal stresses in addition to allowing hydrogen to diffuse to a more uniform distribution throughout the lattice and also to diffuse out of the steel.

Endogenous Nonmetallic Inclusions. In contrast to the nonmetallics picked up from outside the melt itself (exogenous inclusions), another class of inclusions can be produced within a metal in both the liquid and the solid states. Such nonmetallics are often referred to as endogenous or indigenous inclusions because they result from reactions within the metal as temperature and composition change. For example, if the solubility of an element in the host lattice (solvent) is exceeded, precipitation of this species may occur. Combinations of certain elements are also favorable from a chemical or thermodynamic standpoint, and such combinations form endogenous nonmetallic inclusions.

Kiessling has thoroughly documented the nonmetallics that occur in steels.[18] Typically, these include the various amalgamations of sulfides, oxides, nitrides, and hydrides. Other metals and alloys may also contain endogenous nonmetallics. Among those subject to significant inclusion formation are the alloys of aluminum, copper, and titanium and also the high temperature alloys based on nickel and cobalt, the latter group experiencing its own particular potentially harmful, second phase formations — namely, sigma and Laves phases.

Unfortunately, nonmetallic inclusions are an inevitable consequence of alloying and melting-and-casting practices. The metal-producing industry attempts to control them because they can seriously impair both the performance and the reliability of metals in service situations.

The deleterious nature of nonmetallic inclusions depends on several factors, including inclusion type or chemical composition,

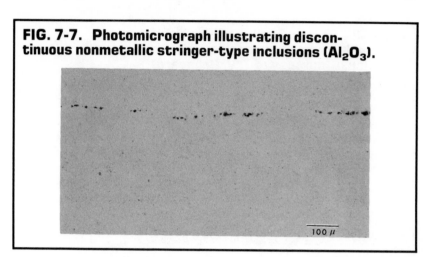

FIG. 7-7. Photomicrograph illustrating discontinuous nonmetallic stringer-type inclusions (Al₂O₃).

volume percentage, shape, orientation with respect to applied stresses, and mechanical properties of the inclusions as compared with the surrounding matrix. Correspondingly, these imperfections can have a significant effect on the mechanical properties and performance of engineering alloys. The following parameters are most seriously affected, especially transverse to the direction of mechanical working:

1. Tensile ductility
2. Impact strength
3. Fatigue properties

Let us briefly examine the effects of nonmetallic inclusions on these parameters.

Tensile Ductility. Many investigators have established relationships between nonmetallic inclusions and a decrease in ductility as measured by percentage reduction of area (transverse) resulting from a tensile test.[19-22] The *inclusion type*, the *quantity* or *volume per cent*, and the *degree of mechanical working* are the primary factors which influence such relationships. Inclusions which are hard and brittle can become fragmented during working and result in discontinuous stringers (Fig. 7-7). Or they may stay intact and cause microcracking in the adjacent metal matrix as shown in Fig. 7-8. Nonmetallics which are softer and readily deform during the working operation result in more continuous stringer-type inclusions (Fig. 7-9). Overall, the presence of nonmetallic inclusions can significantly lower the

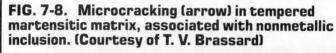

FIG. 7-8. Microcracking (arrow) in tempered martensitic matrix, associated with nonmetallic inclusion. (Courtesy of T. V. Brassard)

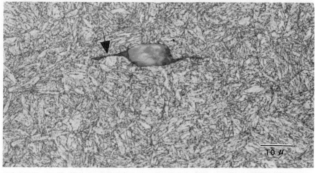

ductility of a metal subjected to tensile loading, an effect that both designers and alloy producers should be aware of.

Impact Strength. The effect of nonmetallic inclusions has been to lower the Charpy impact strength of metals, particularly in the transverse direction.[23] In this situation, the nonmetallics probably act as stress concentrators, thereby promoting the rapid propagation of a crack. The net result is a lowering of the fracture stress or the load necessary to fracture a notched Charpy specimen.

Fatigue Properties. Both the initiation and propagation of a fatigue crack can be affected by nonmetallic inclusions. As pointed out in Chapter 6, the failure mechanism of fatigue not only accounts for a substantial number of premature failures but does so in a very insidious and catastrophic manner. Ordinarily, little or no warning accompanies such a fatigue failure, and the applied stresses are usually far less than the maximum design stresses. Since the fatigue life of any component consists of the cycles required for crack initiation followed by the cycles required to propagate the crack to critical size, any imperfections or defects which assist these processes decrease the safe life of a component or structure.

Several investigators have demonstrated that nonmetallic inclusions can influence fatigue life.[24-32] Although the actual relationships are complex, the following inclusion characteristics apparently affect the fatigue strength or fatigue life of alloys: inclusion content (volume per cent), size, distribution and orientation, and chemical composition of the nonmetallics. The consensus of most investigators is that as inclusion content increases, fatigue properties deteriorate.

FIG. 7-9.   Example of continuous nonmetallic stringers (MnS).

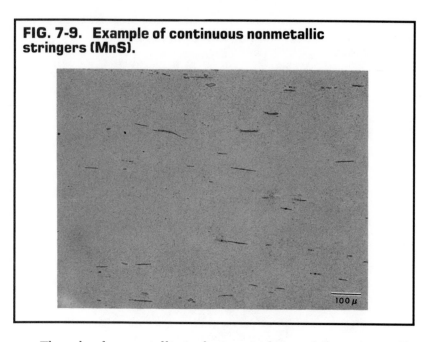

100 μ

The role of nonmetallic inclusions in fatigue failures is vividly demonstrated in the following example of a failed electrical probe. This component fabricated from low alloy steel was cyclically loaded during its operation. Cracks initiated at four separate sites in a radius as shown in Fig. 7-10(a). Examination of the fracture surface revealed nonmetallic inclusions at the origin of the fatigue cracks. A close-up of one initiation site is shown in Fig. 7-10(b). Note the included particle at the origin. Clearly, these inclusions acted as stress raisers during the operation of the component and promoted fatigue crack initiation, thus shortening the life of the part. Such an obvious contribution to the failure of a component or an engineering structure could be interpreted as a significant processing defect in product liability litigation, particularly if such nonmetallic inclusions are abnormal or atypical in similar components and materials.

Nonmetallic inclusions also affect the life and reliability of components which undergo rotation and bending in service. These components principally include shafts and rolls. Figure 7-11 clearly shows the influence of a nonmetallic inclusion on the fatigue failure of a rotating shaft. The fatigue crack, in this instance, initiated at a subsurface particle and grew radially outwards until it reached the surface. Subsequently, the fatigue crack propagated through the shaft until failure occurred.

FIG. 7-10. Fracture of an electrical probe component.

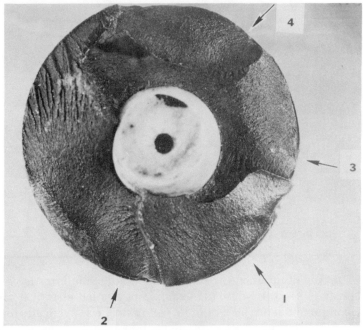

(a) Fracture surface showing four separate crack origins

(b) SEM fractograph showing nonmetallic inclusion at initiation site in crack number 1

FIG. 7-11. Example of nonmetallic inclusion initiating fatigue failure in rotating shaft.

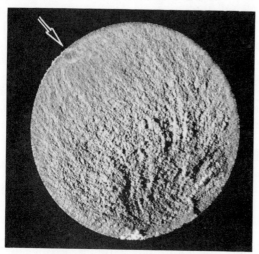

(a) View of fracture (arrow denotes initiation site)

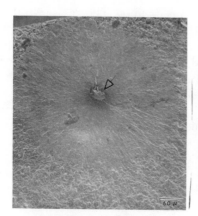

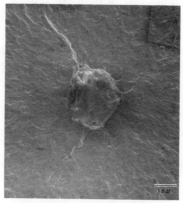

(b) SEM micrograph of initiation site (arrow denotes inclusion)

(c) Close-up of inclusion in situ, revealing its morphology and role as crack initiator

In summary, we can expect nonmetallic inclusions to form in engineering alloys, especially steels. Typically, the formation of endogenous inclusions is difficult to prevent. However, there are methods available to the steelmaker and alloy producer that will minimize and control the occurrence of certain nonmetallics in their product. First, the charge materials used in melting steels should be relatively clean and free from injurious and inclusion-forming elements such as sulfur, phosphorus, and tramp elements.* Second, oxygen and hydrogen should be removed as efficiently as possible. This is accomplished by deoxidation (the addition of deoxidizers and/or vacuum techniques) and degassing. Degassing primarily removes hydrogen by subjecting the liquid alloy to a vacuum (vacuum degassing).

Alternatively, consumable remelting by VAR and ESR can produce alloys with a minimum of nonmetallic inclusions. We alluded to these melting operations earlier in this chapter as a means of alleviating chemical segregation. Coincidently, the formation of certain nonmetallics is diminished because dissolved gases are removed by these processes.

Unfortunately, the addition of extra processing such as vacuum degassing, VAR, or ESR is reflected in the final cost of the alloy. In fact, it is virtually an axiom that the "cleaner" (i.e., having fewer inclusions) an alloy is, the more it must cost. Therefore, the problem becomes one of how clean an alloy needs to be. The answer to this question lies principally in the ultimate application of the material. Since it is widely recognized that nonmetallic inclusions can significantly influence the initiation and growth of cracks in engineering materials and structures, their reduction and elimination should be considered in the production of an alloy, especially when the material will be utilized in components of a critical nature — that is, parts whose premature failure could cause loss of life, injuries, or serious property damage.

The techniques for avoiding deleterious concentrations of nonmetallic inclusions and injurious second phase particles are available to alloy producers. In addition, SAE specifications AMS 2300 and 2301 detail the magnetic particle inspection procedure for aircraft-type quality steel cleanliness, and the ASTM Standard E-45 presents the recommended practice for determining inclusion contents of

---

*If these materials cannot be removed, they should at least be controlled. For example, manganese is added to control the formation of FeS in steels, which causes hot shortness (cracking).

steel by metallographic (microscopic) methods. Furthermore, nonmetallic inclusions can be assessed by the Fairey Inclusion Count system and the J-K method (Swedish Iron Works). These techniques utilize a numbering system to rate the severity of inclusions. However, we must caution against the use of microscopic methods for assessing or specifying inclusion contents. Such an approach necessarily involves examining many small fields, usually at $100 \times$ magnification, to determine a representative inclusion content. This type of examination can be very misleading! For example, suppose nonmetallic inclusions tend to agglomerate in a specific region of a casting or ingot. A random sampling of your component or material may reveal an acceptable level of inclusions — completely missing the deleterious congregation located elsewhere. Remember, microscopic examination is basically a destructive test, involving cutting and polishing. Therefore, it will not be performed on a serviceable component unless a sacrificial coupon is attached. Even then, the question arises: does the coupon material validly represent the bulk material?

Serious consideration must be given to the methods for preventing or minimizing nonmetallic inclusion formation for the following reasons:

1. Routine product quality control
2. Applications where safety and reliability require quality material
3. State-of-the-art techniques are available
4. Reduction of product liability potential

FORMING OPERATIONS

The preponderance of metallic components presently in service have very likely been formed by one of the various metalworking processes. Among the more commercially important methods are forging, rolling, drawing, and extrusion. Generally, most metalworking is performed hot because metals deform more easily at elevated temperatures and therefore less force is required. Since hot working accounts for the bulk of parts and materials that are pre-formed, we will direct our attention to these methods.

The various forming processes differ with respect to the type and manner of applied loads and the shapes that can be produced. Correspondingly, each forming operation potentially has characteristic defects associated with it. In the following section we will attempt to point out the more important defects and their potential effects on product liability. However, the perceptive reader will no doubt detect the similarities in defects and defective conditions that can be produced by the various metalworking processes. Our intent, therefore, is to examine a few processes from their potential product liability standpoint with the belief that the underlying theme can be recognized and applied to the other metalworking processes.

Forging

Forging operations, in addition to producing a desired shape, also result in some benefits for the material. These include closing certain internal imperfections such as gas pores, shrinkage cavities, and voids; refining the dendritic cast structure of an ingot; and improving the mechanical properties of a material in the direction of forging. Such processes may be open-die or closed-die, and they may be performed hot or cold. The interested reader is referred to several comprehensive treatises on forging procedures.[33-35]

Although forging is usually a beneficial practice, certain defects can be produced during this mechanical working operation. The principal defects that can result from forging include *laps, hot tearing, forging bursts, thermal cracks,* and *forging cracks.* Other deleterious conditions which can result from the forging operation are residual tensile stresses, mechanical fibering, and burning. These last three conditions are more subtle than the defects previously mentioned and therefore are not ordinarily observed or detected at the production stage. Yet they pose just as serious a threat to the performance of a forged component as do the others.

First, let us examine three of the more common, detectable defects which can result from forging, both open- and closed-die operations. Then we will discuss the three more subtle defective conditions mentioned above.

Laps. A forging lap, as the term implies, consists of a "folding-in" of the workpiece surface. Such a defect can occur when the forging stock contains sizable surface irregularities — for example, protrusions or gouges. The folded metal forms a fin which does not bond properly because of oxidation on the previously exposed surfaces. The resultant lap can, under the right circumstances, behave like a crack. Examples of forging laps are shown in Fig. 7-12. Note the forged-in scale in (a), evidence of an oxidized surface which has been incorporated into the workpiece, and the acuity of the lap in (b). Both features are potential crack initiators and can readily contribute to

FIG. 7-12. Examples of forging laps from press forging operation.

(a) Cross section showing forged-in scale (dark gray); same scale as (b)

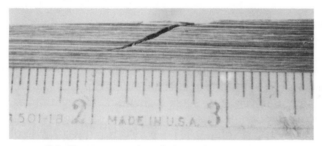

(b) Cross-sectional view showing the notch acuity of the lap (scale in inches)

premature failures, through the various mechanisms that are discussed in Chapter 6.

Forging laps are generally associated with the following conditions:

1. Surface irregularities on the workpiece being forged.
2. The interaction between vertical and horizontal sections of a forging.
3. Nonuniform deformation of the metal as it flows into the die cavity.
4. Excessive die wear, improperly sized forging dies, or improper forging techniques. An example of a forging lap created by upsetting the workpiece is schematically illustrated in Fig. 7-13.

FIG. 7-13. Schematic illustration of a forging lap in a closed-die operation.

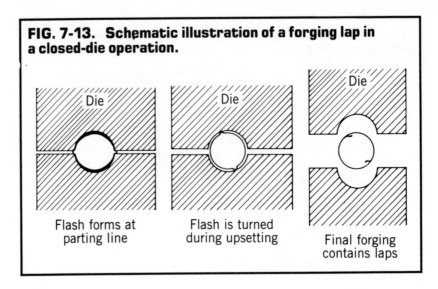

| Flash forms at parting line | Flash is turned during upsetting | Final forging contains laps |

Forging Bursts. This type of metalworking defect consists of an internal rupture of the workpiece. These ruptures can occur if the forging temperature is insufficient for the metal being deformed. The stresses generated by the working operation exceed the elevated temperature strength and cause separations especially along planes or areas of material weakness such as chemical segregation, or low melting or brittle second phases. An example of an internal rupture due to working stresses and planes of weakness in a low alloy steel is shown in Fig. 7-14. In fact, low melting point constituents or segregations may possibly melt at the temperature used to preheat the metal for the forging operation. Such a condition is referred to as "hot shortness" and is usually associated with the formation of FeS in steels. Hot shortness is alleviated in steels by the addition of manganese. However, compounds containing Mn-S-Fe-O can still form under the right conditions and promote hot shortness. Therefore, control of sulfur content is important and its reduction in steel (with the exception of resulfurized steels) is usually beneficial.

Thermal Cracks. This type of metalworking defect results from uneven or nonuniform heating of the workpiece or metal to be forged. This condition can also produce cracking during cooling, following the working operation. Stresses are produced by the uneven expansion or contraction of the metal, along with stresses induced by crystallographic transformations. If these stresses cannot be

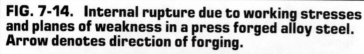

FIG. 7-14. Internal rupture due to working stresses and planes of weakness in a press forged alloy steel. Arrow denotes direction of forging.

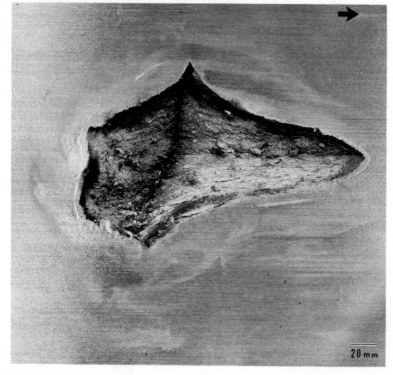

accommodated by the lattice, cracking results. In large pieces and heavy sections, this problem is aggravated because the surface layers heat and expand sooner than the cool interior. An example of thermal cracking in large billets is shown in Fig. 7-15. The cracks occurred in an induction-type preheat process, which produces rapid heating, and resulted in scrapping the billet.

As we previously mentioned, there are other, more subtle, deleterious conditions which can result from forging: namely, *residual stresses, mechanical fibering,* and *burning.* Since these conditions are not readily detected, or for that matter normally anticipated, they pose a potential threat to the integrity of the finished part. Just what are these factors, and how are they produced?

FIG. 7-15. Thermal cracking in a 20 inch diameter forging billet. Cracks extended deep into the workpiece.

Residual Stresses. This type of stress refers to the stress left in a material after a process is completed on the metal. It can result from forging or almost any metalworking process which causes plastic or permanent deformation. The magnitude of residual stress depends on the amount of deformation and the temperature at which it is performed. Typically, residual stresses increase as the amount of deformation increases and the working temperature decreases. Such stress can be beneficial, if it is compressive in nature, but it can be detrimental if tensile, tending to pull the metal lattice apart.

Residual stresses can also develop because of nonuniform deformation through the thickness of a forged part. The situation may be further aggravated by uneven temperature distribution throughout the part.

Generally, the effects of residual tensile stresses are deleterious. They can cause warping and distortion of components, which will be reason for rejection or further processing such as straightening. Worse yet, residual stresses may cause failures. For instance, residual tensile stresses can decrease the applied loads necessary to produce fracture.

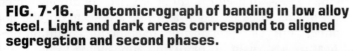

FIG. 7-16. Photomicrograph of banding in low alloy steel. Light and dark areas correspond to aligned segregation and second phases.

500 μ

They can also assist cracking by stress corrosion, hydrogen embrittlement, liquid metal embrittlement, and fatigue. Thus, residual stresses should be prevented or relieved, particularly in components where the above failure mechanisms are possible.

Mechanical Fibering. This is the term given to the alignment of grains, chemical segregation, and second phase particles (inclusions) in the direction of forging. As these constituents of the metal elongate during the working operation, they create planes and areas of weakness in the forged product. Depending upon the concentration and distribution of these factors, plus the amount of forging reduction, the resulting "fibered" product may look and behave like a composite material. The term "banding" has been applied to worked products which exhibit a heavily fibered (aligned) structure. An example of banding in low alloy steel is shown in Fig. 7-16. It may exhibit different properties (mechanical and physical) in different directions. Such behavior is called anisotropic. This condition, anisotropy, was previously discussed in Chapter 6 with respect to the proper utilization of engineering materials. If a forged component exhibiting anistropy is employed in the wrong orientation, the weakest directions of the material may see the greatest applied loads. Such a circumstance is potentially dangerous and invites failure.

Burning. Overheating and burning are two conditions which can occur in metals to be forged when they are heated to the hot working temperature. The term burning refers to the condition produced by heating a steel in an oxidizing atmosphere to the point where fusion or oxidation occurs at the austenitic grain boundaries. An example of burning is shown in Fig. 7-17. These regions subsequently tear and rupture during forging. An alloy that has been burned cannot be salvaged and should be reverted as charge material. On the other hand, overheating a metal to be forged may not necessarily result in scrapping. Slight overheating can be ameliorated by the subsequent forging operation. However, more severe overheating — that is, temperatures which affect the lowest melting point constituents (second phases) — can lower the ductility and toughness of heat-treated forgings.

Rolling

Rolling is another principal method for forming metals. Like forging, it is performed both hot and cold, although the bulk of rolling is accomplished hot. Basically, the operation consists of passing the material to be shaped between two rolls revolving at the same speed in opposite directions. The workpiece to be rolled is heated to the appropriate rolling temperature and is plastically deformed to the desired shape. A typical rolling process is schematically illustrated in Fig. 7-18. Since there are similarities between rolling and forging — i.e., in both processes the workpiece is preheated and then plastically deformed — some of the defects and deleterious material conditions that may occur in rolling are very similar to those in forging. However, one major difference between the two forming processes is that rolled plate and sheet are produced without any upsetting. In other words, the metal is successively deformed in the same direction until the desired thickness is obtained. In contrast, the preponderance of forgings and other rolled products such as structural shapes and merchant bar can be upset — that is, rotated with respect to the shaping forces. Upsetting tends to distribute the deformation more uniformly and reduce the amount of mechanical fibering or banding in a worked product. A detailed treatment of rolling theory and practice is given by Wusatowski.[36]

Rolling Defects. There are numerous defects which may be produced by rolling. The interested reader is referred to a comprehensive narrative by U.S. Steel covering these problems.[37] Therefore,

**FIG. 7-17. Burning in a nickel base high tempera-
ture alloy. Note penetration and oxidation of
grain boundaries.**

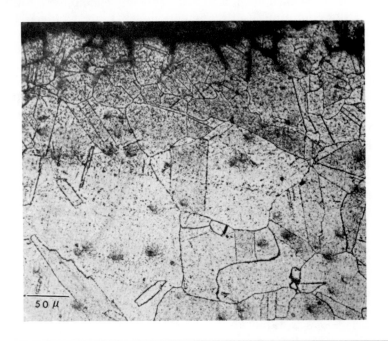

$50\,\mu$

**FIG. 7-18. Schematic of a sequential rolling
process.**

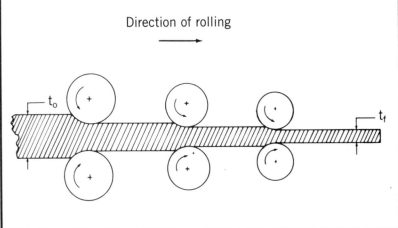

Direction of rolling

we will limit our discussion to some of the more familiar major flaws and material conditions which can contribute to failures, thus affecting the product liability of a producer.

Rolled-in Scale. One of the most frequently encountered rolling defects, this condition results when previously formed surface scale is incorporated into the rolled product. Although this problem is more likely to be associated with semifinished bars, plates, and slabs, rolled-in scale may persist into the finished product. If the defect is not detected and enters service, it can, under the right circumstances. act as a stress concentrator and contribute to crack initiation, thus playing a role in the failure mechanisms discussed in Chapter 6.

Seams. Seams are created by the closure of surface gouges, holes, notches, crevices, etc., during a pass through the rolls. These seams are closed, but not metallurgically joined, because of previous oxidation of their surfaces.

Fins. Fins are protrusions which form when the workpiece is oversized or when lateral spreading has not been allowed for. Less sharp protrusions are called overfills, a more frequent occurrence than fins.

Underfills. Just the reverse of overfills and fins, underfills occur when the work is rolled insufficiently in one dimension.

Laps. As in forging, laps can be produced in rolling when protrusions such as fins and overfills are subsequently rolled into the surface of a workpiece.

Pipe. This term refers to the defect which develops when piping is still present from the ingot stage. The formation of pipe in the ingot was discussed earlier in this chapter under the section Ingot Defects. During rolling, this condition is exacerbated by the working, which tends to elongate the piped region, producing an internal cylindrical cavity.

The aforementioned defects are just some of the flaws and imperfections that can develop during rolling processes. In addition we wish to point out some conditions associated with rolling which also can affect the quality and reliability of the final product. These conditions are not obvious defects and therefore can more readily escape detection.

Banding. Rolling processes can inherently create heavy mechanical fibering or banding, depending on the alloy, the condition of the original ingot, and the amount of rolling deformation (reduction). An example of banding in rolled, low carbon steel plate is shown in Fig. 7-19. Such fibering produces anisotropic behavior in mechanical and physical properties — in other words, different properties in different directions.

FIG. 7-19. Banding in rolled, low carbon steel plate. (Courtesy of F. A. Heiser)

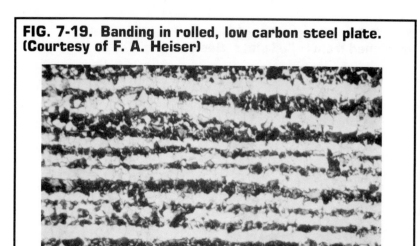

The anisotropic conditions mentioned in Chapter 6 can lead to sudden, unanticipated failures, especially in rolled products. For example, a brittle failure occurred in a locking collar device for a hydraulic system. On investigation, we discovered that the component had been machined from rolled plate, as illustrated in Fig. 7-20. The compact tension "test bars" denote the orientation of crack planes with respect to the rolling direction (RD) as per ASTM Standard E-399 (Plane-Strain Fracture Toughness Testing of Metallic Materials). Note that the actual fractures occurred in the S-T and S-L directions. The composite photomicrograph (Fig. 7-21) shows that these crack planes (fractures) are parallel to the banding in this plate and therefore the component was actually loaded in its weakest structural orientation. In fact, mechanical property evaluation showed that the S-T direction exhibited an average ductility of just 4% RA while the L-T direction exhibited an average ductility of 57% RA for comparable yield strength levels.

The point we are making in this example is that anisotropy of mechanical properties was not considered in the production of this component. The loads applied to the locking collar in service produced tensile stresses on the weakest planes in the material. Clearly, this is not the most desirable structural situation, yet many failures can be attributed to this condition. A principal reason is perhaps that

FIG. 7-20. Illustration of failure of a locking collar machined from rolled plate. See also Fig. 7-21.

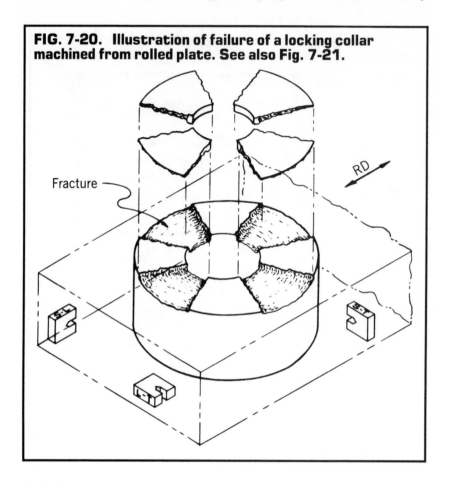

Fracture

RD

the anisotropic behavior or response of a heavily fibered structure is not apparent or appreciated by the manufacturers and processors of such components. Nevertheless, it does result in a material condition that can seriously influence the performance and safety of engineering components and structures, as this example reveals.

Banded structures are not only anisotropic with respect to strength, ductility, and toughness but also exhibit poor crack growth resistance along the interface between banded constituents. Such a condition can be conductive to a significantly reduced fatigue life and a lower resistance to delayed failure by environmentally assisted mechanisms such as stress corrosion cracking. An example of preferential cracking associated with banding is shown in Fig. 7-22.

Residual stresses can result from nonuniform cooling following a rolling operation, and this condition may be aggravated by the shape

FIG. 7-21. Composite photomicrograph from locking collar (Fig. 7-20), showing orientation effects of rolled structure.

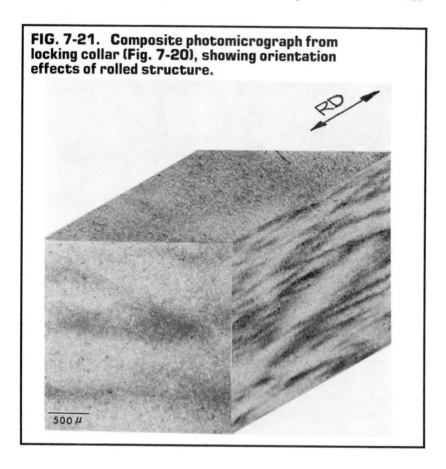

of the rolled section. Consequently, warping or distortion may occur. Worse yet, the product may not receive further treatment and the built-in residual stresses will not be relieved. As discussed in the section on forging, this condition is usually harmful and promotes premature failures. The insidious aspect of these stresses is that they are unanticipated and of unknown magnitude and location. They are ordinarily unaccounted for in design and therefore have the potential to produce, or at least contribute significantly to, unexpected failures.

Drawing

Wire-drawing processes constitute another means of shaping metals. Briefly, the procedure consists of hot rolling wire rod, then cold drawing this shape through dies until the desired diameter is obtained. An example of the drawing process is shown schematically

FIG. 7-22. Photomicrograph depicting preferential fatigue crack growth associated with banding. (Courtesy of F. A. Heiser)

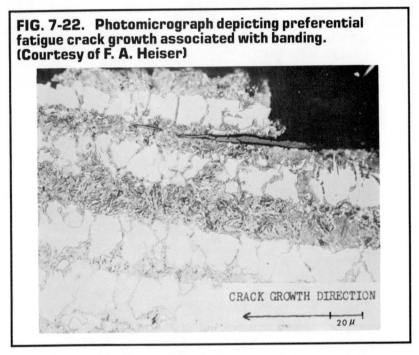

CRACK GROWTH DIRECTION

20 μ

in Fig. 7-23. The first portion of this process, rolling the rod or slender bar, can have certain defects associated with it. Not unexpectedly, they are identical to the rolling defects previously discussed in this chapter — for example, pipe, seams, and slivers (fragments of metal rolled into the surface). Defects produced at this stage can obviously end up in the drawn wire. The principal defects that can develop in drawn wire are categorized as follows:[38]

1. Internal defects
2. Surface defects
3. Improper mechanical properties

Internal Defects. In much the same fashion as the other deformation processes, piping and chemical segregation are the prevalent defects that occur in drawn wire. These conditions cause uneven heat treatment and brittleness. They can also be responsible for delamination or peeling in wire products.

Surface Defects. This category of defects includes scratches, nicks, seams, and slivers. They result from inadequate or improper

FIG. 7-23. Schematic illustration of drawing operation.

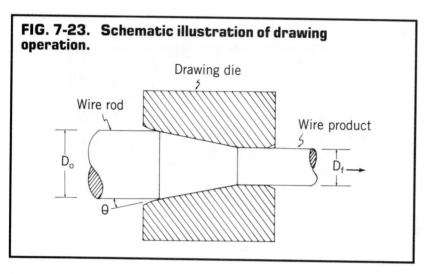

lubrication, poor die condition, or contamination of the rod being drawn. Slivers, as previously mentioned, can be incorporated into the surface of the metal as it is drawn through the dies. Since there is no metallurgical bond between the slivers and the wire, they can act as notches and promote cracking under some circumstances. Seams also act as longitudinal stress raisers in wire products.

Mechanical Properties. The properties of drawn wire may vary considerably, depending on the alloy composition, the reduction (cold working) it has undergone, and any subsequent heat treatments. Exact specifications and requirements are necessary for the proper, safe applications of wire products. In addition, real material variations such as segregation of alloy within the wire can produce appreciable variations in its final mechanical properties. Such variations, as we pointed out in the section on materials selection and specification in Chapter 5, may inadvertently result in material below a minimum acceptable property level. In the event that such variance has not been considered or compensated for by design, unexpected failures are invited.

Residual Stresses. Drawing processes that are conducted at lower temperatures tend to produce residual stresses. Although we have discussed residual stress in the section on forging, it is fundamental to most deformation processes. For instance, take the case of brass or bronze that has been cold worked by drawing, extrusion, or

FIG. 7-24. Season-cracked brass connector.

rolling. If residual stresses from such operations are not relieved, they tend to promote intergranular cracking in service environments that are corrosive. This type of failure is referred to as *season cracking*, a form of stress corrosion cracking. One of the particularly harmful characteristics of season cracking is that it is spontaneous and can occur with no warning.

An example of season cracking with catastrophic consequences is illustrated in Fig. 7-24. In this particular case, a brass connector on a natural gas line season-cracked in the presence of ammonia ions to the point of failure and allowed gas to escape. The escaping gas eventually accumulated and exploded, causing severe fire damage to a structure.

The dangers of season cracking can be minimized by careful control of the cold work processes and by a subsequent thermal stress relief treatment in the neighborhood of 260-425 °C (500-800 °F).

Again, we emphasize the potentially injurious effects of producing residual stresses in forming operations and not relieving them prior to service. Often such a deleterious material condition goes unnoticed until it is too late. The fabricator of a component should be aware of whether the processing method introduces residual stresses and what the consequences are relative to service life.

Extrusion

The process of extrusion consists basically of enclosing a preheated blank of metal in a container with a die located at one

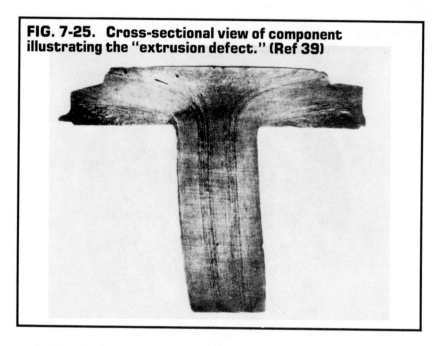

FIG. 7-25. Cross-sectional view of component illustrating the "extrusion defect." (Ref 39)

end. The die has an opening of the desired product cross section. Pressure is applied to the heated blank to force it through the die until the metal plastically flows into the desired shape. Certain defects have been specifically associated with the extrusion process and we will refer to one type which is representative of this operation.

The Extrusion Defect. This term refers to a characteristic extrusion flaw similar to piping which has long been associated with copper alloys.[39] This defect originates when the oxidized surface at the rear corners of the billet is enfolded during extrusion. The path of this defect is shown in Fig. 7-25, for an extruded bar product. Severe cases of the extrusion defect can result in separation between the core and the outer annulus of a rod as illustrated in Fig. 7-26.

This defect and the incorporation of any impurities into the extruded product can seriously affect the quality and performance of the material. The points to be emphasized are as follows:

1. Relatively sound, defect-free billet stock can eventually contain a serious flaw after extrusion.
2. This type of defect is internal and therefore not readily detected.
3. Nondestructive inspection methods, viz., ultrasonics and radiography, can be utilized for detecting such flaws and minimizing their consequences in service.

FIG. 7-26. Annular separation of extruded billet due to severe extrusion defect. (Ref 39)

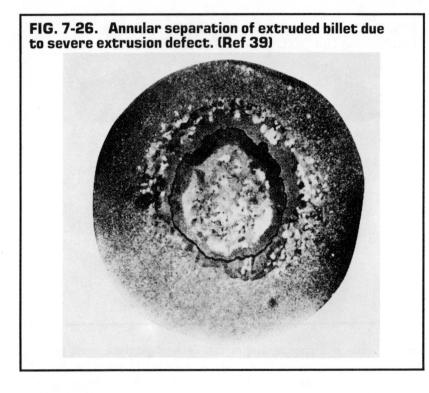

Summary

In summarizing this section on metalforming processes, we would like to point out that although final mechanical properties are usually developed through heat treatment (with the exception of cold worked alloys), the metalforming parameters and procedures are instrumental in providing a sound, defect-free component that will respond properly to heat treatment. The defects and harmful material conditions we have discussed are essentially preventable by careful forming practice, assuming that one has a reasonably sound, defect-free material to start with. Additionally, most of the forming defects discussed above and the other clearly obvious ones we chose to omit are detectable by one of the commercially available nondestructive testing (NDT) techniques such as magnetic particle inspection, liquid dye penetrant, X-ray radiography, ultrasonics, and eddy current inspection.

NDT methods and equipment have progressed to the stage where they are quite accurate and reliable for detecting and identifying most metalworking defects. Portable units are also available for some techniques, including radiography, ultrasonics, and eddy current, thus allowing on-site inspection of many products in the mill or shop. Hence, metalforming manufacturers and producers should utilize NDT to its fullest extent as a quality control measure. Not only does such a procedure guard against defective products being placed into service, it also provides valuable feedback to the processing engineer, perhaps signaling potential problems in an operation or clues to the origin of an existing processing problem.

From the product liability position, detection and elimination of potentially harmful defects resulting from metalworking operations is a necessity. In some instances, specifications and requirements dictate the standards which a component or material must meet. These requisites may or may not clearly spell out the permissibility of defects. However, we must emphasize that metalforming defects and the deleterious conditions that can result from any forming process may ultimately be held as *the reason* for a catastrophic failure or perhaps as a primary contributor towards such a failure. The inevitable questions will arise sooner or later: could such a defect or harmful condition have been prevented? Could the defect have been detected and eliminated? The processor of a component should strive to be aware of the end use of the component and what degree of defect is acceptable.

We present this line of thought for the personnel who are responsible for the fabrication and forming of metals and the control of their quality. What position would you rather be in, answering these questions in your shop or plant or responding to such queries in a courtroom?

REFERENCES

1. Weinstein, A. S., Twerski, A. D., Piehler, H. R., and Donaher, W. A., *Products Liability and the Reasonably Safe Product*, John Wiley and Sons, New York, 1978, p. 17.
2. Ibid., p. 31.
3. *Metals Handbook*, 8th Ed., Vol. 10, "Failure Analysis and Prevention," American Society for Metals, Metals Park, Ohio 44073, 1975, p. 315.
4. Briggs, C. W., *The Metallurgy of Steel Castings*, McGraw-Hill Book Co., New York, 1946.
5. *The Solidification of Steel Castings*, C. W. Briggs, Ed., Steel Founders' Society of America, 606 Terminal Tower, Cleveland, Ohio 44113, 1965.
6. *Metals Handbook*, 8th Ed., Vol. 11, "Nondestructive Inspection and Quality Control," American Society for Metals, Metals Park, Ohio 44073, 1976, p. 139.

7. Chalmers, B., *Principles of Solidification*, John Wiley and Sons, New York, 1964, p. 126.
8. Hollomon, J. H., and Jaffe, L. D., *Ferrous Metallurgical Design*, John Wiley and Sons, New York, 1947, p. 16.
9. Duckworth, W. E., and Hoyle, G., *Electroslag Refining*, Chapman and Hall, London, 1969.
10. Bunshak, R. F., Ed., *Vacuum Metallurgy*, Reinhold Publishing Corp., New York, 1958.
11. Krutenat, R. C., Ed., *Vacuum Metallurgy*, Science Press, Princeton, 1977.
12. Irving, R. R., "VAR Materials Are Back in the Saddle Again," *Iron Age*, Oct. 1, 1979.
13. Prengamon, M. E., "VAR and ESR: Do They Measure Up?," Watervliet Arsenal Technical Report, WVT-TN-75048, Aug. 1975.
14. "The Vacuum Treatment of Molten Nonferrous Metals, Including Vacuum Distillation," Section 3, Proc. 5th Int. Conf. on Vacuum Metallurgy and ESR Processes, Munich, Germany, Oct. 1976, pp. 75-104.
15. *The Making, Shaping and Treating of Steel*, United States Steel Corp., Pittsburgh, Pa., 1964, p. 548.
16. Ibid., p. 847.
17. Ibid., p. 1079.
18. Kiessling, R., and Lange, N., *Non-Metallic Inclusions in Steel*, Parts 1-4, The Metals Society, London, 1978.
19. Paliwoda, E. J., "Mechanical Working of Steel-I," AIME Metallurgical Conferences, 21, 1964, p. 126.
20. Rauson, J. T., and Mehl, R. F., *Proc. ASTM* 52, 1952, p. 779.
21. Welchner, J., and Hildorf, W. G., *Trans. ASM* 42, 1950, p. 455.
22. Carter, C. J., Cellitti, R. A., and Abar, J. W., "Effects of Inclusions as Measured by Ultrasonic Methods on the Mechanical Properties of Aircraft Quality Steel," AFML-TR-68-303, Jan. 1969.
23. Kinzel, A. B., and Crafts, W., *Trans. AIME* 95, 1931, p. 143.
24. Pelloux, R. M. N., *Trans. ASM* 57, 1964, p. 511.
25. Heiser, F. A., and Hertzberg, R. W., *J. Basic Engr.* 93, No. 2, 1971, p. 211.
26. Volchok, I. P., Kovchik, S. E., Panasyuh, V. V., and Shulte, U. A., FKHMM (Soviet Materials Science) 34, 1967, p. 439.
27. Stulen, F. B., Cummings, H. N., and Schulte, W. C., Proc. Int. Conf. on Fatigue of Metals, 1965, p. 439.
28. Cummings, H. N., Stulen, F. B., and Schulte, W. C., *Trans. ASM* 49, 1957, p. 482.
29. Ransom, J. T., *Trans. ASTM* 46, 1954, p. 1254.
30. Cummings, H. N., Stulen, F. B., and Schulte, W. C., *Proc. ASTM* 58, 1958, p. 505.
31. Atkinson, M., *J. Iron Steel Inst.* 195, 1960, p. 64.
32. Fisher, J. J., and Sheehan, J. P., "Effect of Metallurgical Variables on the Fatigue Strength of AISI 4340 Steel," Armour Research Foundation, WADC-TR-58-289, Feb. 1959.
33. Sabroff, A. M., Boulger, F. W., and Henning, H. J., *Forging Materials and Practices*, Reinhold Book Corp., New York, 1968.
34. *Forging Design Handbook*, P. M. Unterweiser, Ed., American Society for Metals, Metals Park, Ohio 44073, 1971.

35. *Forging Industry Handbook*, J. E. Jenson, Ed., Forging Industry Association, Cleveland, Ohio, 1966.
36. Wusatowski, Z., *Fundamentals of Rolling*, Pergamon Press, Wydawnictwo, "Slask," Katowice, 1969.
37. Ref 15, p. 757.
38. Ref 15, p. 799.
39. Pearson, C. E., and Redvers, P. N., *The Extrusion of Metals*, John Wiley and Sons, New York, 1961, p. 157.

Secondary Manufacturing Processes That Affect Product Liability

Certain products require further processing after the primary stages of fabrication. The details of these primary stages are discussed in Chapter 7. The secondary processing may include such operations as further forming (hot or cold), heat treatment, surface hardening, metallurgical joining (welding), electrochemical plating, and machining. This additional processing may eventually increase the potential for defects and defective material conditions. In other words, the more processing an item gets, the more likely it is that something will go wrong, with the process and/or the material.

In this chapter, we will discuss some of the problems associated with the secondary manufacturing processes and relate these problems to the product liability potential of a component, or perhaps more appropriately, the product liability position of the company that performed the operation(s).

HEAT TREATMENT

Heat treatment is routinely applied to a very large percentage of engineering alloys in order to improve certain mechanical properties. These metals include both ferrous and nonferrous, although the ferrous alloys undergo the bulk of heat treatment because of their unique response to heat treatment — namely, hardening. It will be advantageous to discuss these two groups separately, because of the inherent differences in the materials themselves and the different manner in which they are heat treated. Furthermore, as we discuss the various aspects of heat treatment processes, their influence on the quality and performance of materials will become evident. Again, we will attempt to point out the problem areas that can result in defects and harmful material conditions which in turn affect the product liability of the processor or manufacturer.

Introduction to Heat Treatment

First, a brief introduction to the terms and processes involved in heat treatment is in order. Such a prelude will assist the reader who is somewhat unfamiliar with the actual heat treatment procedures.

In the case of ferrous alloys or steels, a typical heat treatment may consist of the following operations:

Full Annealing — An elevated temperature treatment conducted above the upper critical temperature for hypoeutectoid steels and above the lower critical temperature for hypereutectoid steels, followed by very slow cooling (in the furnace) to room temperature. The purpose of this treatment is to soften the material, make it more machinable, and relieve residual stresses. Also, since this is an elevated temperature process, diffusion can take place, reducing chemical inhomogeneity. A near equilibrium structure is produced during this process, as shown in Fig. 8-1.

Normalizing — This treatment is similar to annealing, except that the cooling sequence is accomplished in air (preferably outside the furnace). The purpose of this treatment is also to soften the steel, improve chemical homogeneity, and relieve residual stresses.

Hardening — This procedure consists of heating an alloy to its austenitizing temperature until the carbides have dissolved. Then the material is cooled sufficiently fast to form a martensitic or fully hardened structure.

Tempering — A hardened alloy is heated to below the lower critical temperature (to prevent recrystallization). This process allows carbides to form in the steel, thereby increasing the ductility and toughness of the hardened material. The treatment also serves to relieve residual stresses produced by the rapid cooling in the previous hardening operation. The quenched-and-tempered structure is shown in Fig. 8-2.

FIG. 8-1. Microstructure of carbon steel in the annealed condition. Etched in 2% nital. (Courtesy of P. J. DeMeo)

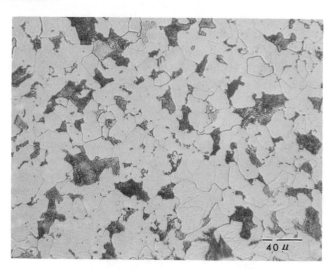

FIG. 8-2. Microstructure of carbon steel (same as in Fig. 8-1) in the quenched-and-tempered condition. 2% nital. (Courtesy of P. J. DeMeo)

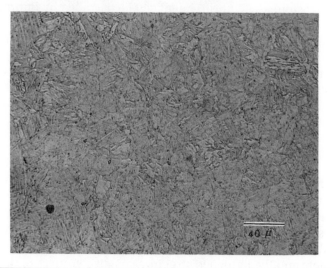

There are several other types of specific thermal treatments performed on steels such as process annealing, spheroidize annealing,
and stress relieving. However, these topics will not be included in
this discussion.

The second group of metals we will include in this section on
heat treatment consists of nonferrous alloys such as aluminum, magnesium, titanium, and the nickel or cobalt base superalloys. Heat
treatments performed on these materials may include *annealing*, to
soften a cold worked (strain hardened) alloy; *solution treatment* at the
proper temperature to dissolve second phases, followed by rapid
cooling to retain the solid solution; and finally, *aging* or *precipitation
hardening*, which causes the gradual precipitation of fine submicroscopic second phase particles. The latter process results in the
strengthening of certain nonferrous alloys.

Heat Treatment of Steel

First let us discuss some of the problems that can occur in connection with the heat treatment of steels, problems that may eventually contribute to or influence the failure of a component or structure
and in turn affect the product liability of the processor. Improper
or inadequate heat treatment can lower mechanical properties —
yield and tensile strength, ductility, and impact toughness, for
example — and can decrease the fatigue and fracture toughness characteristics of an alloy. Also, the resistance of a steel to corrosion,
stress corrosion, and embrittlement can be degraded by improper or
inadequate heat treatment.

For example, if an alloy is annealed or normalized at temperatures above the normal range for that material, excessive grain growth
may take place, especially if grain growth inhibitors such as aluminum, molybdenum, vanadium, etc., are absent. Even though the steel
may get a subsequent austenitizing treatment, the excessively large
grains that were previously produced can negatively affect mechanical properties. Very often, this degradation in properties cannot be
altered by any later thermal treatments.

Hardening. Occasionally a component or material is incompletely austenitized during the hardening operation. That is, the
temperature or time does not reach the point where recrystallization
and dissolution of carbides are sufficient. When this condition occurs, subsequent transformation to martensite, upon cooling, is also
incomplete and the hardened structure contains vestiges of the start-

ing microstructure. Such incomplete transformation produces hard and soft areas in the structure and manifests itself in both lower yield strength and impact toughness. Comparison of an incompletely austenitized structure and a sufficiently austenitized one is shown in Fig. 8-3.

On the other hand, excessive austenitization can also have deleterious consequences. As previously stated, if temperatures above the normal range for a specific heat treatment procedure are experi-

FIG. 8-3. Examples of heat treated low alloy steel microstructures.

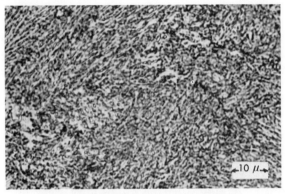

(a) Incompletely austenitized; note quasi-eutectoid appearance

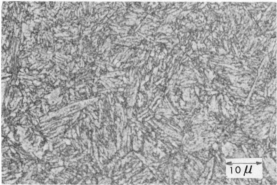

(b) Sufficiently austenitized — structure completely transformed

enced, enlargement of the grains results. Such a condition can also occur if prolonged austenitizing times are used. The result is an impairment of the impact toughness of certain alloys, particularly at low service or operating temperatures.

Slack Quench. Even when a material is properly austenitized, the rate of cooling (quenching) must be sufficiently fast to transform the austenite completely to martensite, usually the desired product. However, several factors affect quenching — the quenching medium, its temperature, agitation of the quenchant and the workpiece, etc. If all these factors are not well controlled, an insufficient quench, often referred to as a *slack quench*, can occur. The result is transformation to products (microstructures) other than martensite. The mixed microstructure which results does not give the best combination of strength and toughness and can also decrease fatigue life.[1] Unfortunately, like incomplete austenitizing, slack quenching or incomplete transformation to martensite is not readily detectable. Except for a relatively low value of the ratio of yield strength to ultimate tensile strength (YS/UTS), this condition can go unnoticed in many alloys. With an inadequate degree of inspection, many components and engineering materials could get into service without actually meeting design requirements and therefore possibly promote premature failures.

Quench Cracking. Just the opposite of slack or insufficient quenching, a cooling rate which is too severe for a particular alloy can produce cracks. These cracks are commonly referred to as quench cracks and result when the stresses produced by thermal gradients and crystallographic transformation exceed the fracture stress of the material. Ordinarily, quench cracking is influenced by the size and geometry of the piece, quench severity, alloy chemistry, and both metallurgical and mechanical stress concentrators.

When quench cracks are severe, it may be necessary to scrap the component. Although this is certainly an undesirable economic situation, potentially dangerous defects are actually being screened out of service. When detected in the manufacturing stage, such flaws may possibly be removed by a succeeding machining operation. Also, the process itself can be examined to find the causes of such cracking.

However, when quench cracks are not readily apparent and escape detection, they can provide convenient sites for crack initiation, thus promoting failure by any number of mechanisms. The following example briefly illustrates the consequences of processing defects, including a quench crack, on the performance of a thick-wall steel pressure vessel.

A thick-wall hydraulic cylinder failed abruptly during pressurization. The origin of the failure was easily identified, as shown in Fig. 8-4. This defect contained exogenous nonmetallic included material which subsequently contributed to a quench crack during heat treatment. Together, these defects constituted a severe flaw and caused the premature fracture of the component.

FIG. 8-4. Quench crack in a thick-wall pressure vessel.

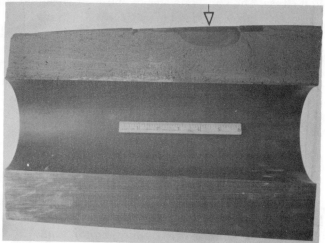

(a) Crack (arrow) served as initiating defect for failure

(b) Close-up of fracture shows two distinguishable pre-cracks, 1 and 2; note dark gray color indicating oxide produced during tempering

Residual Stresses. We previously alluded to the formation of residual stresses in the discussion of mechanical working. Similarly, heat treatment can produce residual stresses in a material. Generally, residual compressive stresses are beneficial, while residual tensile stresses are harmful.

The quenching operation produces the stresses, both thermal and transformational, which can cause residual stress. Nonuniform cooling further complicates the situation. If the residual stresses are not relieved, the hardened component may be in a very undesirable state of residual stress. The consequences of "built-in" residual stress are demonstrated by the warping schematically illustrated in Fig. 8.5.

In order to minimize the effects of residual stress and alleviate cracking, the hardened piece should be tempered immediately. If this is not possible, a temporary "holding temper" in the vicinity of 350 °F is recommended to relieve residual stresses. The tempering process causes the martensitic structure to contract. However, if heating during this operation is not uniform, distortion or cracking can occur.

Residual stress conditions which escape detection can seriously affect the performance and safety of a component or structure in service. Residual tensile stresses lower the magnitude of applied

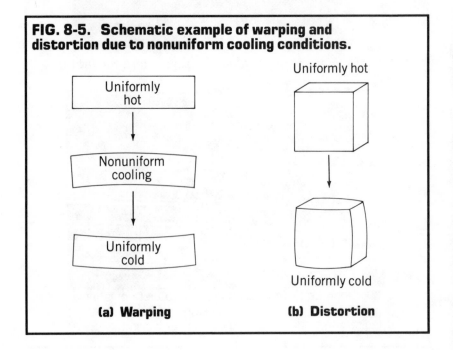

FIG. 8-5. Schematic example of warping and distortion due to nonuniform cooling conditions.

Uniformly hot

Uniformly
hot

↓

Nonuniform
cooling

↓

Uniformly
cold

Uniformly cold

(a) Warping   **(b) Distortion**

stress (working load) that may be necessary to cause yielding or even fracture. These same stresses also promote crack initiation and growth by other failure mechanisms such as fatigue, creep, stress corrosion cracking, hydrogen embrittlement, and liquid metal embrittlement.[2,3] Since these failure mechanisms operate under the influence of tensile stresses, any addition to those resulting from the external forces will assist crack propagation and hasten premature, perhaps even catastrophic, failure.

An example of the influence of both quench cracking and residual tensile stress on fatigue failures in gear teeth was reported by Gaydos.[4] In this particular situation, the residual tensile stresses from heat treatment were oriented or operating in the same direction as the applied cyclic stress. The cracked gear teeth are shown in Fig. 8-6. As a result of this condition, the service life of the gear was drastically shortened. Gear teeth which have undergone similar cracking are illustrated in Fig. 8-7.

FIG. 8-6. Cross section of gear teeth which failed from quench cracks (inset). Arrows denote cracks. (Ref 4)

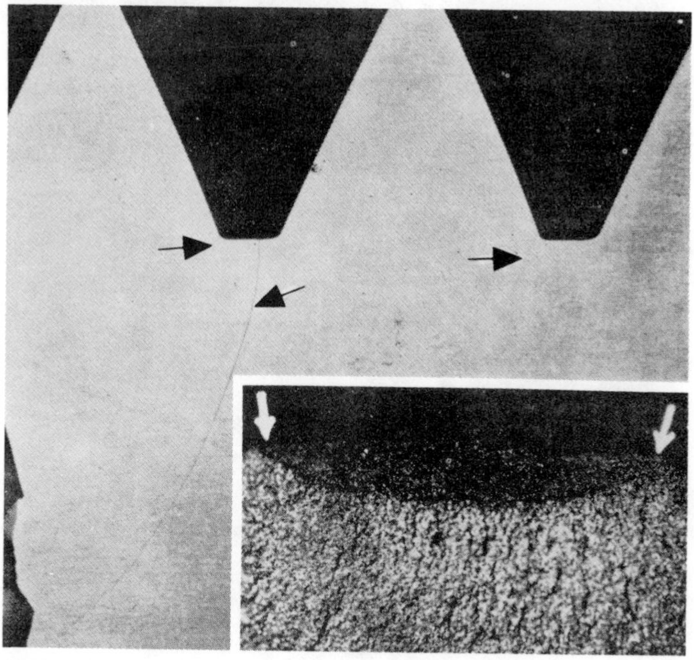

FIG. 8-7. Cracking in gear teeth.

(a) SEM micrograph of cracking
(arrow); note missing tooth

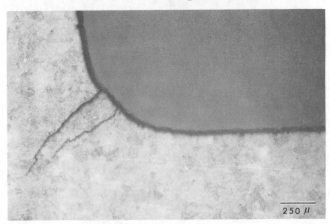

(b) Photomicrograph showing
depth of cracking from tooth root
radius; etched in 2% nital

Insufficient Tempering. After quenching to a hardened martensitic structure, a steel should be tempered to relieve harmful residual stresses and develop the required mechanical properties. If the material receives no temper or is tempered inadequately, its sensitivity to stress concentrators (notches, scratches, nonmetallic inclusions, etc.) is intensified. Also, the susceptibility of a component to brittle failure increases.

An example of such a deleterious material condition is illustrated by the failure of the 4130 steel knob shown in Fig. 8-8(a). A metallographic examination revealed the microstructure to be essentially untempered martensite (Fig. 8-8b) with an R_C hardness of 46, substantially higher than the required range R_C 35-41. A fatigue crack initiated in the root of a thread (Fig. 8-8c) and grew to critical size. Brittle fracture, as shown in Fig. 8-8(d), ensued.

Overtempering. While overtempering may be caused in some cases by heating at the correct temperature for excessive periods of time, tempering at or near the recrystallization temperature of an alloy can result in a serious loss of strength. Many alloys exhibit a precipitous drop in yield and tensile strength with only slight increases in temperature beyond the proper tempering range.

Temper Embrittlement. Tempering certain steels in the general range of 450-540 °C (850-1000 °F) can produce temper embrittlement. The effect of this condition is to lower the fracture stress at small strains, thereby inducing brittle failures. Temper embrittlement is also manifested in an increase in the impact transition temperature.[5] Recall from Chapter 6 that increasing the ductile-to-brittle transition temperature promotes brittle fracture. Since this type of failure usually gives very little warning, the results can often be catastrophic.

Decarburization. As the term implies, decarburization is a depletion of the surface carbon content of a component. It occurs during thermal treatment in uncontrolled atmospheres[6] and can also occur at high service temperatures. Since many heat treatment operations are performed in furnaces with access to ambient air, decarburization is a common problem.

The result of decarburization in steels is a relatively soft, ferritic microstructure adjacent to the exterior surface or, for that matter, adjacent to the surfaces of any existing cracks or discontinuities in the surface of the part. When decarburization is excessive or

FIG. 8-8. Microstructurally induced failure in a 4130 steel knob.

(a) Knob assembly; arrow denotes fracture in threaded portion

(b) Photomicrograph showing untempered martensitic structure

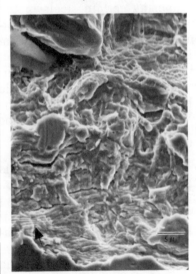

(c) SEM fractograph of initiation site; arrow denotes fatigue striations

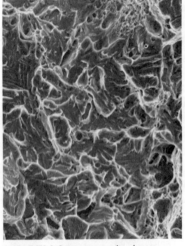

(d) SEM fractograph showing brittle nature of ensuing failure (quasi-cleavage model). Same magnification as (c).

deep, the weakened "case" can reduce fatigue life by promoting crack initiation.

The basic problem is that a decarburized condition is not anticipated in the design. Therefore, if the producer does not make allowances for preventing this condition or removing it by subsequent machining, a structurally deficient component can enter service and possibly contribute to a premature failure.

Heat Treatment of Nonferrous Alloys

As mentioned earlier in this section, heat treatment of nonferrous alloys consists chiefly of *annealing, solution treatment,* and *aging.* Rather than discuss the processing problems associated with the individual groups of alloys and these specific treatments, we propose to point out several general defects or defective conditions which are common to all heat treatable nonferrous alloys. Based upon the legal information and definitions presented earlier, a processor can weigh the consequences of producing such conditions in his material or product and take appropriate action.

Age Hardening

Since aging or precipitation hardening is the final stage in heat treating nonferrous alloys, it governs the final mechanical properties a component exhibits. For example, an aluminum alloy containing 4.5% copper will form a fine, submicroscopic precipitate of $CuAl_2$ when aged artificially at temperatures above 300 °F. This precipitate strengthens the material significantly, resulting in yield strength increases on the order of 200%. From the designer's point of view, this increase in strength is very desirable. He can now design with a material which has a relatively high strength-to-weight ratio.

However, this type of alloy and certain titanium alloys will undergo a noticeable decrease in strength as aging temperature or aging time increases. Figure 8-9 demonstrates the effects of *overaging* on 2014-T4 aluminum. Overaging causes the precipitate particles to grow and eventually coalesce, resulting in a degradation of the yield and ultimate tensile strength of the alloy. Process temperatures and time, in addition to chemical composition, are therefore very important factors to control in producing such materials and subsequent components. Undoubtedly the designer using these alloys will expect them to be in the optimum physical and mechanical condition. The performance and reliability of components produced from precipitation-hardened alloys will be extremely sensitive to processing.

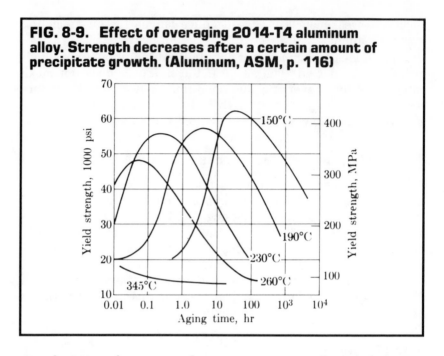

FIG. 8-9. Effect of overaging 2014-T4 aluminum alloy. Strength decreases after a certain amount of precipitate growth. (Aluminum, ASM, p. 116)

Any deviations from proper heat treatment can influence the integrity of the components and hence the product liability position of the producer.

SURFACE HARDENING

Certain engineering components benefit from an abrasion-resistant surface, especially those that are placed in moving contact situations. Such parts include bearings, gears, wheels, shafts, cams, valves, etc. A number of methods have been developed to increase the hardness and thus the strength of these sliding or contact surfaces. The commercial processes include carburizing, nitriding, induction hardening, flame hardening, shot peening, and case hardening by laser or electron beams.

Although the individual processes differ, they all basically produce a component with a hard case and a relatively softer core. For instance, carburizing and nitriding increase the surface concentration of carbon and nitrogen to produce increased hardness. Induction hardening, like flame hardening and beam hardening, produces a crystallographic transformation at the surface and results in a hard, untempered martensite case on the component. Shot peening is analogous to cold working or strain hardening the surface layers, thus producing a harder or stronger surface by localized plastic deformation.

As we have pointed out, many components require case hardening for reasonable performance and life. Moreover, another reason exists for case hardening certain engineering components. The surface-hardening processes may produce a state of residual compressive stress in the case, thereby increasing the fatigue life of the structure. For components that are subjected to cyclic loading, this can be an extra bonus. Unquestionably, case hardening can be very beneficial for some engineering materials.

Why then do we include surface or case hardening in our discussion of processing factors which can affect the product liability of a producer? Primarily, the reason is that surface-hardened components have been frequently involved in failures, and the origin of these failures has been traced to the case-hardened region. A detailed analysis of this kind of failure is presented by Ebert, Krotine, and Troiano.[7] It will be sufficient to say that the problem develops at the *case-core interface* of surface-hardened components where a triaxial stress state can exist. This situation creates a notch-like effect at the interface and promotes crack initiation. Therefore, a tough core is vital because the characteristics of the core dominate at the interface. When stresses exceed the elastic limit of the core, a complex stress situation forms at the interface and the potential for fracture increases.

The depth and uniformity of case are important factors in determining the fracture resistance of surface-hardened parts. Figure 8-10 shows a failure in a gear between the teeth where the case depth was very shallow because of nonuniform induction hardening. This is a good example of where the *cure* for abrasion was *unhealthy* for the rest of the part!

In addition to giving proper consideration to case hardening from a design standpoint, the processor or manufacturer must exercise diligent process control. Case depth and hardness must be uniform if defective conditions are to be minimized and avoided.

WELDING

Generically, welding is a process that involves joining metals with a metallurgical bond — that is, establishing attractive atomic forces between the metals to be joined. Welding may be conducted by pressure methods, which employ deformation and heating to achieve

FIG. 8-10. Example of failure in case-hardened gear teeth due to nonuniform hardening. (Ref 32, p. 269)

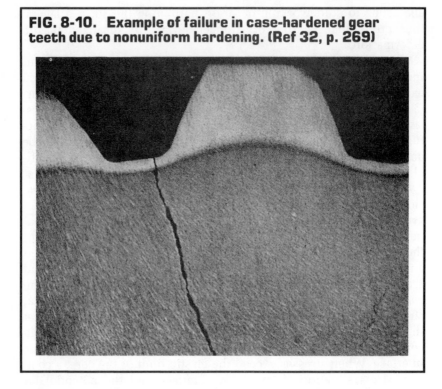

the metallic bond, or by *fusion* methods, which consist of melting and solidification to produce bonding.

Savage has classified the major types of welding defects as shown in Table 8-1.[8] In addition, the causes and prevention of weld defects have been thoroughly documented in the literature along with illustrations of several common defects.[9-12] We will further comment on some of these imperfections with respect to the welding process, especially the defects which have a high incidence rate and are not easily detected.

Pressure Welding

Pressure welding must employ intimate metal-to-metal contact in order to develop sound welds. The surfaces to be metallurgically joined have to be free from contaminants such as dirt, grease, oil, paint, etc., which may degrade or embrittle the weld. Another form of contamination is the oxide film that inevitably forms on most metallic surfaces. If this film is not removed just prior to the welding operation, it will prevent intimate contact and can also subsequently contaminate the weld.

An example of the effects of sulfur contamination on pressure-welded surfaces is shown in Fig. 8-11.[13] The embrittling effects are obvious and certainly promote premature failure of the weld or base metals in service.

Interference Weakening. This imperfection refers to a condition of residual tensile stresses produced during pressure welding which can cause localized brittle failure along the plane of the weld interface. As we have consistently emphasized, residual stresses are particularly dangerous because they are not readily detected by commercial NDT techniques. A post-weld thermal treatment designed to relieve stresses will benefit components that are prone to this defect.

Delaminations. As the term implies, delaminations are material separations oriented parallel to the direction of working in rolled products such as sheet and plate. In the pressure-welding process

TABLE 8-1. Welding Defects (Ref 8)

1. Defects involving inadequate bonding
   - A. Interface weakening in pressure welds
   - B. Incomplete fusion or penetration in fusion welds

2. Foreign inclusions
   - A. Oxide films in fusion welds
   - B. Slag inclusions
   - C. Delaminations
   - D. Tungsten inclusions in GTA welds

3. Geometric defects
   - A. Undercutting
   - B. Excessive reinforcement

4. Metallurgical defects
   - A. Defects related to microsegregation
     - (1) Hot cracking and microfissures
     - (2) Cold cracking and delayed cracking
     - (3) Stress-relief cracking
     - (4) Strain-age cracking
     - (5) Gas porosity
   - B. Problems arising from metallurgical reactions
     - (1) Embrittlement
     - (2) Structural notches

FIG. 8-11. Sulfur embrittlement of the root bend in Nickel 200 sheet. (Courtesy of Huntington Alloys, Inc.)

Left side of joint was cleaned with solvent and a clean cloth before welding. Right side, wiped with solvent and a dirty cloth, exhibits extensive cracking.

these separations can be reoriented parallel to the weld interface resulting in an inferior weld which can fail in a brittle manner. If the weld is nondestructively inspected when a separation actually exists, detection of this condition is not difficult. However, if the delamination is partially or weakly bonded, detection is extremely difficult.

In order to produce strong, reliable pressure welds, processors must adhere to the following criteria:

Use diligence in cleaning the surfaces to be welded.
Carefully control the application of pressure.
Carefully control the temperature of the welding operation.
Utilize post-welding stress relief, if appropriate.
Critically inspect the weld zone by NDT techniques, depending upon the application, for surface discontinuities and internal flaws.

Fusion Welding

The fusion-welding process consists of actual melting and solidification of the metals to be joined. Usually a filler metal of comparable chemistry is added to the weld. Since this operation depends on melting and solidification of both filler and base metals, fusion welds can and do experience, on a smaller scale, the same maladies as castings and ingots. Furthermore, the metal adjacent to the fusion zone, commonly referred to as the heat-affected zone (HAZ), undergoes what amounts to heat treatment as the heat from the weld puddle is dissipated into the surrounding joined materials. This condition is schematically illustrated in Fig. 8-12, which depicts the cross section of a butt-welded joint and the metallurgical ramifications that accompany it.[14]

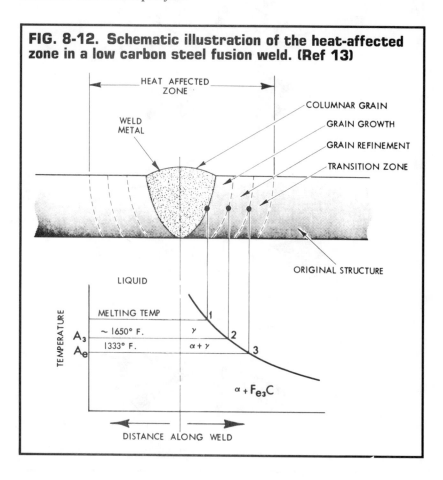

FIG. 8-12. Schematic illustration of the heat-affected zone in a low carbon steel fusion weld. (Ref 13)

Incomplete Fusion or Penetration. As classified in Table 8-1, incomplete fusion or penetration results in a deficient bond, usually located at the root of the weld. Examples of a root gap resulting from incomplete fusion and six other common weld defects are illustrated in Fig. 8-13. A description of unacceptable fillet-weld and butt-weld profiles given by the AWS Structural Welding Code[15]

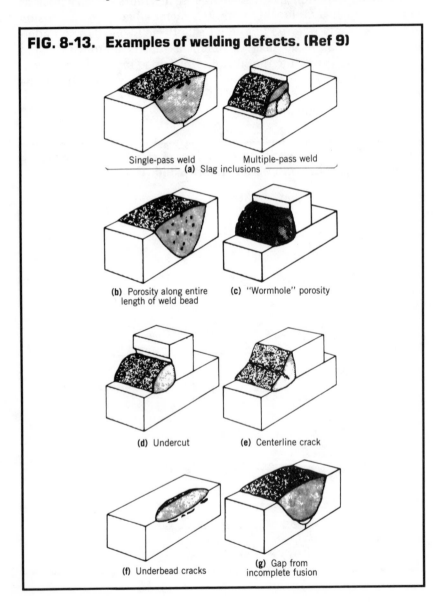

FIG. 8-13. Examples of welding defects. (Ref 9)

Single-pass weld   Multiple-pass weld
(a) Slag inclusions

(b) Porosity along entire length of weld bead   **(c)** "Wormhole" porosity

(d) Undercut   **(e)** Centerline crack

(f) Underbead cracks   **(g)** Gap from incomplete fusion

is shown in Fig. 8-14. Incomplete fusion or penetration reduces the strength of a weld because the contact area is reduced. In addition, these flaws can act as stress concentrators, thus assisting crack initiation and premature failure.

Inclusions. Entrapment of exogenous inclusions and the formation of second phase particles in the weld are analogous to the situations presented earlier in connection with melting and solidi-

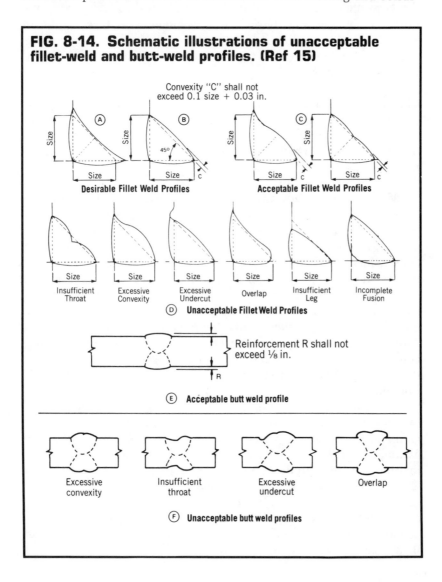

FIG. 8-14. Schematic illustrations of unacceptable fillet-weld and butt-weld profiles. (Ref 15)

fication. Regardless of their source nonmetallic inclusions, depending on their size, shape, orientation, and physical properties, can act as stress raisers. In this capacity, they can seriously degrade the fatigue strength and toughness of the welded component. In some cases — for instance, oxide films — the imperfections are exceedingly difficult to detect.

Geometric Defects. Undesirable angles between the weld and the base metal can be produced by undercutting or by excessive reinforcement (see Fig. 8-14). The consequence may be a significant reduction in fatigue life. It is interesting to note that undercutting, although often ignored, has the worst record for causing mechanical failures in weldments.[16]

For instance, the incident cited in the section on fatigue in Chapter 6 (see Fig. 6-10), involving an amusement-park ride, was directly attributable to undercutting of the fusion weld. The resulting geometric notch served as a stress raiser, promoting initiation of a fatigue crack which eventually caused the failure of the structure with serious consequences to the riders.

Metallurgical Defects. Several defects associated with the fusion welding process can be viewed simply as metallurgical imperfections. They are primarily related to microsegregation of the alloying elements in the fusion zone and in the adjacent heat-affected zone. Since these defects are adequately described in the literature,[17-18] it is sufficient to point out here that they are a consequence of the solidification mechanisms operating in the weld and the solid-state reaction occurring in the base metal immediately adjacent to the area of the weld. In other words, the metallurgical defects listed in Table 8-1 can routinely occur in welds — hot cracking, microfissures, cold cracking, strain-age cracking, porosity, etc. In order to alleviate them, Savage[19] has suggested reducing microsegregation by controlling the welding variables, including the welding velocity, the temperature gradient in the liquid, and the shape of the weld puddle. Consideration should also be given to the super-heat-melt-back region and the region of partial melting, both of which are typically assumed to be part of the HAZ but in fact can undergo melting. Such melting generally produces microsegregation, which is likely to affect the toughness and fatigue strength of the material.

Metallurgical Reactions. Certain steels, such as the low alloy constructional grades and carbon steels containing more than about 0.18% carbon, form martensite when cooled at a sufficient rate. This microstructure is hard and brittle and may crack after welding. When hardenable steels are welded, precautions should be taken to avoid transformation to martensite. Since cooling rate is a major factor, slow cooling rates should be employed. One method of accomplishing slow cooling is to preheat the metals to be joined to a temperature of approximately 90-200 °C (200-400 °F).[20] Subsequently, a slower cooling rate assists transformation to structures other than martensite and reduces the hardness of the weld region. This effect is schematically illustrated in Fig. 8-15.

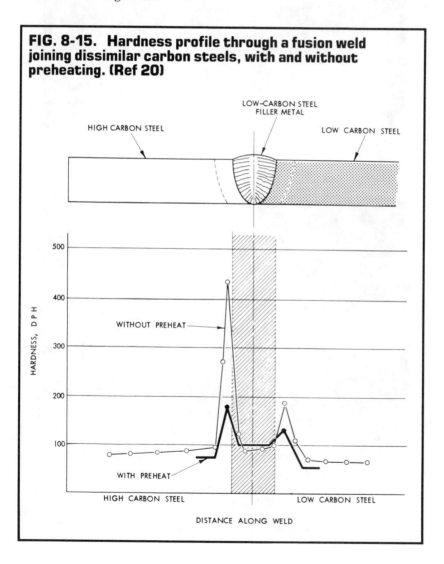

FIG. 8-15. Hardness profile through a fusion weld joining dissimilar carbon steels, with and without preheating. (Ref 20)

FIG. 8-16. Failure in a cast steel cylinder.

(a) Extensive cracking which originated at weld zone (arrow); scale in inches

(b) Close-up of fracture showing quench crack (arrows) in base metal below weld fillet

The effects of rapid cooling on a weld are exemplified by the failure of the cast steel cylinder shown in Fig. 8-16(a). Quench cracks resulted in the material adjacent to the weld (Fig. 8-16b) and led to the overall failure of the cylinder.

Postheating. Postheating a welded joint also tends to reduce the propensity for cracking associated with welding. Postheating treatments are usually employed to obtain mechanical properties, to relieve residual stresses, to avoid hydrogen cracking, and to achieve dimensional stability. The actual time and temperatures used depend on the desired properties and the materials involved.

Residual Stresses. Since the welding process in effect heat treats the base metal adjacent to the weld zone (HAZ), thermal gradients can result in residual stress in and near the weld. The residual stress distribution for a typical butt weld is schematically illustrated in Fig. 8-17. Note that tensile stresses exist in the weld and adjacent material. As we previously pointed out, residual tensile stresses are unfavorable and generally promote crack initiation.

Weld Repair. The process of repairing parts that contain surface imperfections and certain other flaws is widely practiced in many industries, especially the casting industry. This practice actually consists of removing the subject imperfection, usually by grinding or machining, and then filling this region with weld metal. Such a procedure can salvage many components which otherwise would surely be rejected and probably scrapped.

However, weld repair must not be treated lightly or indiscriminately. As we previously cautioned, the weld process inherently alters the structure of a material and can produce defects identical to those produced by melting and solidification. A recent article in Iron Age points out many problems associated with welding repairs.[21] Depending on the alloy content of the component being repaired, preheating and postheating may be beneficial, if not absolutely necessary. The weld preparation and filler metal must be compatible with the component. In other words, the repair procedure must be conducted as carefully as an original welding procedure.

Naumann and Spies point out the importance of this point in connection with the failure of a rear wheel of a sports car on an express highway.[22] Their investigation disclosed that the fracture was caused by stresses and cracking generated during weld repair of the suspension bushing. In their conclusion they state:

This event very clearly demonstrates the lack of responsibility on the part of machine shops that execute or permit repair-welds on highly stressed machine parts and especially vehicular components.

These circumstances bring up a point that we will briefly address here. The original manufacturer of the suspension bushing component may have had absolutely no knowledge of the weld repair process conducted on the part. In fact, many components may receive alteration and modification subsequent to their original manufacture. Such changes may be produced by further forming or shaping, heat treatment, welding, etc. Yet if one of these altered

FIG. 8-17. Schematic illustration of residual stresses in a typical butt-welded plate.

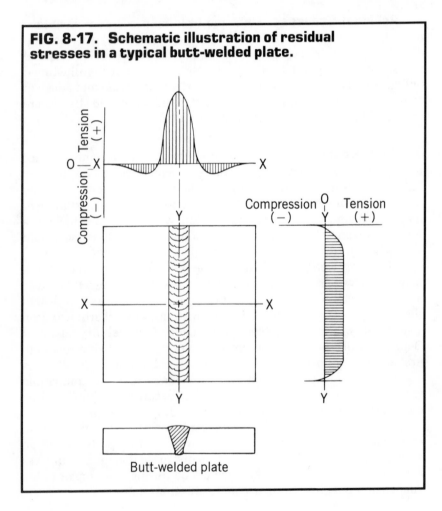

Butt-welded plate

components is involved in a product liability suit, the original manufacturer may well be placed in the position of defending his product. Therefore, in addition to good design and workmanship, we emphasize the following points for minimizing product liability:

1. Thorough inspection and quality control measures.
2. Accurate record keeping and documentation of production, processing, and inspection-testing.
3. Careful examination of the subject component, if it becomes necessary, to disclose any changes introduced since original manufacture and acceptance.
4. Comparison with comparable items previously produced or currently being manufactured if appropriate. Such comparison should include properties and structure.
5. Manufacturers should attempt to ascertain the subsequent processing, if any, that their product will receive. We realize this is not always possible, because the product may change hands through various processors, but since these processes can influence the product liability of the entire production chain, including the original manufacturer, such information is very important.

Summary

Admittedly, there are a host of welding defects, both geometric and metallurgical, that we have not included in this discussion. Indeed, there are probably more types of potential defects associated with welding than with any other single materials process. Of the 112 cases cited in the ASM publication *Case Histories in Failure Analysis*,[23] approximately 20 are associated in some way with welding processes. We could not hope to adequately cover them all. The interested reader is referred to the references cited in this section, which do a comprehensive job of identifying and discussing these processing imperfections and flaws.

The main points in our discussion are as follows:

1. When the welding defects are commonplace and readily identifiable, manufacturers and processors must exercise adequate quality control procedures to detect and exclude such flaws from service. The tools for inspection and identification of harmful defects are commercially available. In other words, insufficient detection and identification is a poor defense when a welding flaw is responsible for, or contributes to, a failure.

2. When the more elusive but still potentially harmful conditions can be anticipated in welding certain materials or components, the welds should be designed to minimize the resultant harmful or deleterious effects. Such conditions may include hydrogen embrittlement, microsegregation, residual tensile stresses, inclusions, etc. Weld design which integrates both mechanical or geometrical and metallurgical considerations will assist the production of defect-free, better quality welds. Thus the product liability potential of the welding processor and the subsequent chain of processors and marketers is reduced.

MACHINING

Machining processes consist basically of *metal cutting* and *metal removal*. The operations that perform these functions include turning, boring, milling, drilling, and grinding. During machining processes metal chips are produced, and the condition of these chips — that is, their shape and color — is an indicator of how efficient the operation is. A comprehensive treatise on machining processes is given in Volume 3 of the 8th edition of *Metals Handbook*.[24]

Since metal removal by machining processes is necessary to the fabrication and production of virtually all metallic components and structures, machining often tends to be taken for granted. However, the fact that chip formation and removal can involve very high, localized forces and high temperatures should serve to caution the processor or manufacturer. Internal metallurgical changes and reactions may be produced, and these alterations could result in defective products.

Deformation and Strain Hardening

During certain machining operations, such as milling and turning, plastic deformation is produced in both the chip and in the workpiece itself ahead of the chip. Depending on such factors as the cutting speed, the tool angle (rake), the strain-hardening characteristics of the material, etc., a layer of deformed metal may be left in the surface of the workpiece. Such a condition is shown in Fig. 8-18, illustrating that slower cutting speeds can produce deeper zones of deformation. This deformed layer can be considerably harder than the underlying matrix and may subsequently be a problem in service. Such a strain-hardened layer would be less resistant to crack initia-

FIG. 8-18. Photomicrographs showing the effect of various cutting speeds on the microstructure of 4140 steel. Note that faster speeds produce less deformation. (Courtesy of Waterbury Farrel Division — Textron Inc.)

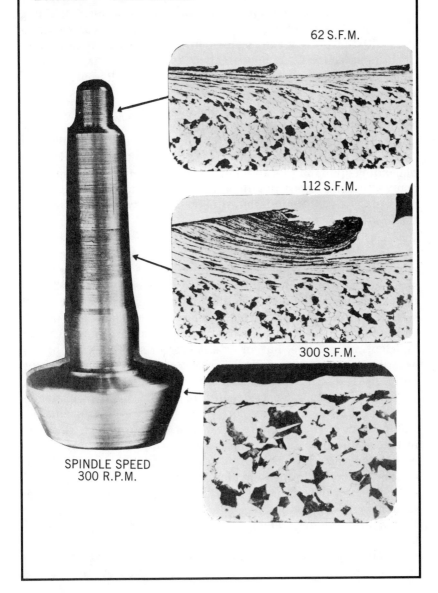

62 S.F.M.

112 S.F.M.

300 S.F.M.

SPINDLE SPEED
300 R.P.M.

tion, especially in the presence of stress concentrators, both surface and subsurface — for example, notches and inclusions. This surface condition can result in decreased fatigue life in components subjected to cyclic loading and also is more sensitive to corrosive attack in aggressive environments.

Tool Marks and Gouges. Since many of the various machining operations involve cutting and chip removal, any obstacle that interferes with this process can leave asperities on the machined surface. Such imperfections, commonly called "tool marks," may appear as gouges, scratches, grooves, etc. These flaws may result from the use of broken or dull tools or from improper machining practice. A number of pertinent examples relating machining defects to premature failures have been presented by VanderVoort.[25] Whenever surface flaws are present in engineering components which are subjected to applied stresses, they can act as stress concentrators. In fact, depending upon their location, they may aggravate an existing stress concentration present in fillets, sharp radii, holes, thread roots, etc. This is a very undesirable situation as we will demonstrate in the following example.

A 4140 steel shaft which operated in a piston-like fashion is shown in Fig. 8-19(a). The shaft failed at the point where it screws into an eye-bolt fixture. Figure 8-19(b) shows the fracture surface of the shaft and clearly displays the beach marks usually associated with fatigue failure. A detailed examination of the threads at the failed section revealed a considerable number of gouges and tool marks, shown in Fig. 8-19(c). This is a case where the material itself was not deficient or defective, but an external factor such as a machining asperity in conjunction with the stress-concentrating effects of the radius in a thread root promoted premature failure of the component.

A similar situation occasionally arises from engraving or stamping identification marks on components. Electric instruments are employed to etch an identifying code on many products. In some instances this process may cause a microstructural change in the underlying material, making it more susceptible to crack initiation.[26,27] Stamped codes and identification marks can also produce a stress raiser, thus assisting crack initiation. The deleterious consequences these markings present to the component and perhaps the overall structure or assembly depend on their location and depth.

Grinding. Grinding may produce surface cracking in hardened steels and tools. Improper grinding practices, including failure to use

FIG. 8-19.   Failure in a 4140 steel shaft.

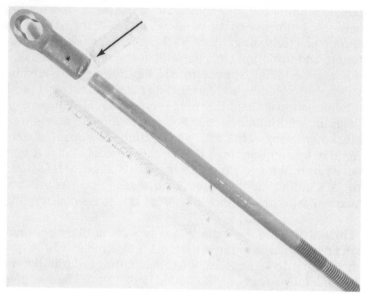

(a) Arrow denotes fracture at the
eye bolt); scale in inches

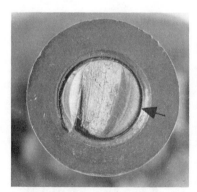

(b) Close-up of fracture
surface showing fatigue
markings; arrow shows
shaft threaded into
eye bolt

(c) Close-up of thread
region and fracture; note
machining asperities in
the threads (arrows)

sufficient coolant, excessive feed rates, and use of incorrect abrasives, can generate stresses in the surface of the workpiece that result in cracking. Grinding can also contribute to incipient cracking in electrodeposited protective coatings, as we will illustrate in the following section on electroplating. Such cracking provides acute stress concentrators and initiation sites for fatigue and corrosive attack.

Regardless of the surface finish, machining and finishing by abrasives, cutting tools, and burnishing operations always produces some deformation or distortion of the surface layers in the workpiece. How significant this deformation is depends upon the material and its eventual applications. In cases where fatigue is a threat, electropolishing may be appropriate.[28] During this process, metal is actually *dissolved* from the surface rather than mechanically removed. The result is a surface with virtually no mechanical disturbances or geometrical asperities to promote stress and stress concentrations.

Thread Rolling. Essentially, the process of thread rolling consists of pressing hardened steel dies into the periphery of a cylindrical blank.[29] Primarily performed cold, thread rolling re-forms the surface of the blank into threads, as illustrated in Fig. 8-20. The dies displace

FIG. 8-20. Schematic example of thread-rolling process. (Ref 29)

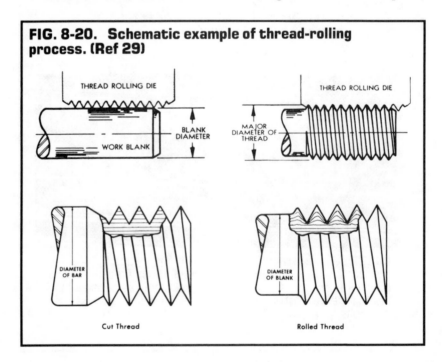

THREAD ROLLING DIE

BLANK DIAMETER

WORK BLANK

THREAD ROLLING DIE

MAJOR DIAMETER OF THREAD

DIAMETER OF BAR

DIAMETER OF BLANK

Cut Thread

Rolled Thread

the blank material to form the roots of the threads and force the displaced material radially outward to form the crests of the threads.

Although some materials can develop folds in the crests of rolled threads, this is not generally a problem because the roots of the threads are more highly stressed. The surface condition of rolled threads is better than that of machined threads, because no cutting action is involved in the thread-rolling process. This alleviates surface imperfections such as tool marks, gouges, tears, etc., which can act as stress concentrators and facilitate crack initiation. The obvious benefit is an improvement in fatigue life, because initiation of cracks in the thread roots is impeded.

ELECTROPLATING

Electroplating is the electrochemical process of depositing one metal upon the surface of another for corrosion protection, abrasion resistance, or appearance. Among the commercially important electroplating metals are copper, chromium, nickel, tin, cadmium, and silver. Since the details of the major plating operations have been thoroughly described,[30,31] we will not discuss the specifics here except to call attention to certain plating processes that have a high potential for producing defects and deleterious material conditions.

Chromium Plating

The electrodeposition of chromium has been widely used for corrosion resistance and appearance. Yet hard chromium is inherently brittle and cracks rather easily. This situation may be further aggravated if the surface of the base metal is in a highly stressed condition, such as that which could result from grinding. An example of "crazing," a network of cracks in electroplated chromium, is shown in Fig. 8-21. These intrinsic cracks in the plate can lead to selective corrosion (pitting) and also contribute to fatigue crack initiation in the underlying material.[32] Both conditions are injurious to the performance and reliability of engineering components and can easily lead to a premature failure with serious consequences.

Hydrogen Charging

During the electrodeposition of chromium and other metals such as cadmium, hydrogen evolves through the reduction reaction which

FIG. 8-21. **Severe cracking of electrodeposited chromium on alloy steel.**

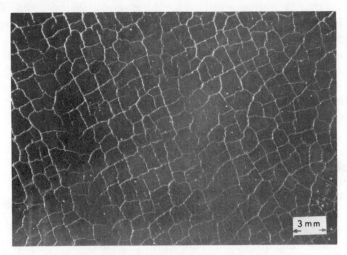

(a) Crack network on surface of chromium plate

(b) Photomicrograph showing cracking through the chromium plate to the base metal

takes place at the cathode. Correspondingly, the more inefficient a plating process is, the more hydrogen evolves. The profusion of hydrogen in some plating operations can result in significant absorption of this element by the base metal. Subsequently, the plated material becomes embrittled. This is particularly troublesome in high strength alloys and in components which contain residual tensile stresses in their surfaces. Recall from Chapter 6 that the criteria for cracking by hydrogen degradation include a susceptible material, applied and/or residual tensile stress, and hydrogen available from dissolved gases or from a reduction reaction such as plating.

An example of failure from hydrogen embrittlement induced by electroplating cadmium on steel screws is illustrated in Fig. 8-22. These particular alloy steel screws received cadmium plating (specification QQ-P-416) for corrosion resistance. During the subsequent failure examination, it was discovered that the cracking initiated in stress concentrators such as the Nylok™ insert hole, the roots of the threaded portion, and the hexagonal cavity in the head. The microstructure was acceptable, but considerable intergranular separation (crazing) was observed in the steel matrix adjacent to the cadmium plate (Fig. 8-22b). Inspection of the fracture surfaces by SEM disclosed an intergranular failure mode indicative of hydrogen embrittlement (Fig. 8-22c).

The failure of these screws was particularly troublesome because of the large numbers involved and the circumstances. For example, frequently the screws cracked severely simply on being tightened in an assembly. Then they separated prematurely after being in service for very short periods of time.

In this case, the evidence showed that hydrogen embrittlement, resulting from the cadmium-plating operation, caused premature failures of the steel screws. It was also noted that specification QQ-P-416 includes a post-plating thermal treatment to relieve absorbed hydrogen. Furthermore, it is common practice to "bake out" high strength steel components after plating, in the range of 175-200 °C (350-400 °F), to facilitate the removal of hydrogen.[33]

SUMMARY

No single text could cover every aspect of the manufacturing processes that can influence the product liability of a producer. Our intention has been rather to point out the subtle or insidious defects and deleterious consequences of certain processes, particularly those

FIG. 8-22.   Failure in cadmium-plated steel screws as a result of hydrogen embrittlement.

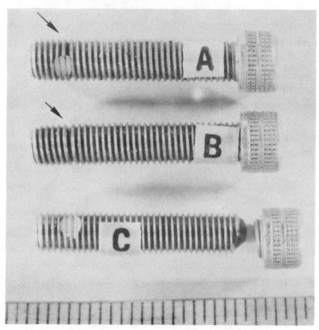

(a) Screws, with cracks indicated by arrows; scale in sixteenths of an inch

(b) Photomicrograph showing extensive crazing (arrow) in the steel matrix next to the cadmium plate.

(c) SEM fractograph of the screws showing brittle failure mode (intergranular fracture) Originally at 920×.

in wide use, and then demonstrate the relationship of these defects to product liability.

There are a number of excellent treatments covering the various aspects of product failures and we have alluded to many of these references in our discussions of both design factors and processing factors. For the interested reader, a list of references on failure analysis is included at the end of this chapter.

In actual practice, the process engineer, designer, specialist, or any person responsible for manufacturing and processing can refer to the legal criteria presented earlier. These criteria will allow one to make a tentative judgment whether a suspected flaw, imperfection, or defective condition constitutes a product liability problem. If the circumstances and consequences of a specific processing flaw are not clear, it is advisable to enlist the aid of legal counsel for a more accurate assessment. Such counsel may be available from a staff lawyer with working knowledge of the product or from an out-of-house attorney expert in matters of product liability. Regardless of where the legal advice is obtained, it will be more beneficial if the counselor is well versed in the intricacies of product liability law.

Review of Liability Theories

At this point, it may be helpful to review the various theories that can be utilized by a plaintiff in a product liability suit. From the litigation standpoint, defects that are produced during processing and manufacturing may fall into the product liability categories shown in Fig. 8-23. Schematically, this chart shows that in certain cases, even if the manufacturer performs a good inspection, he may be tried under the theory of strict liability and the warranty theories for a defective product. If his quality control procedures are inadequate or nonexistent, his product liability exposure increases by the potential inclusion of the negligence aspect. While the details of these theories are presented in Chapter 2, their application to a product liability case may be more evident now that we have discussed a number of actual defects and the processes from which they originated.

Compounding this problem of liability is the fact that not only the original manufacturer but the sub-manufacturers in a product cycle can be included in a product liability suit. For instance, the discovery of a manufacturing defect in a component failure may implicate anyone who has handled that part. Such a product cycle or production chain is illustrated in Fig. 8-24, for a hypothetical component.

Depending on the circumstances, this situation may lend itself to the "apportionment of blame" principle, discussed in Chapter 2, of assigning liability to various members of the chain. The example in Chapter 5 regarding the settlement of a case involving an automotive fan blade incident illustrates this principle.

Since the theory of blame apportionment works in either direction, i.e., up or down the production chain, it behooves each member involved to exercise *product liability prevention*. In other words, the manufacturer of the initial material should protect himself by produc-

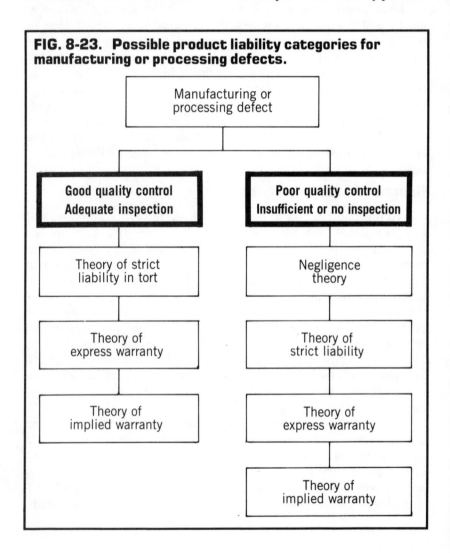

FIG. 8-23. Possible product liability categories for manufacturing or processing defects.

ing a product that not only meets the specifications imposed upon it but also is safe and reliable. He can accomplish this end by good design, careful manufacture, and thorough inspection. Furthermore, to limit his liability exposure, he should take steps to prevent any misuse or misapplication and improper processing of his material by subsequent processors. These are the sub-manufacturers depicted in Fig. 8-24. Sometimes, incorrect processing or use can be prevented by warning statements or labels or even through an educational program for customers.

It should also be incumbent on the sub-manufacturers and processors to ascertain whether the operation they intend to perform on a material is proper and not potentially harmful to the eventual safety and performance of a component. For instance, can the material be shaped under certain conditions? Can it be

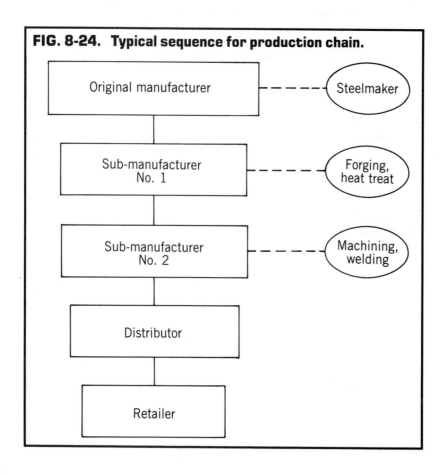

FIG. 8-24. Typical sequence for production chain.

heat treated? Can it be welded? What are the best methods for these processes? Maintaining close communication and technical liaison between members of the production chain will clearly serve to reduce the product liability potential of many components and their associated industries.

Finally, we must stress that full cooperation between the *mechanical designer*, the *process engineer*, and the *product liability counselor* is the key to avoiding serious product liability problems. Unfortunately, our experience is that countless times such cooperation and communication just never exists. Time after time, the designer or developer comes to the materials engineer with a failed component only to receive the terse response, "Why in heaven's name did you make it from that alloy?" Or the product engineer goes to the process engineer with a failure only to hear, "Why did you specify that heat treatment? We could have told you it wouldn't work!"

Surely the most opportune time for collaboration between the responsible parties is before a product liability suit arises, not during a scramble to put together a defense after being named in a litigation. As the life cycle of a product progresses from mere conception to final marketing, continual interaction between the designers, the processing engineers, and the product liability advisors will reduce the threat of product liability problems to a company. Such cooperation is also very likely to improve the overall quality of a product with an attendant improvement in performance and reliability.

BIBLIOGRAPHY ON FAILURE ANALYSIS

Wulpi, D. J., *How Components Fail*, American Society for Metals, Metals Park, Ohio, 1966.

Source Book in Failure Analysis, American Society for Metals, Metals Park, Ohio, 1974.

Case Histories in Failure Analysis, American Society for Metals, Metals Park, Ohio, 1979.

Metals Handbook, 8th Ed., Vol. 9, "Fractography and Atlas of Fractographs," American Society for Metals, Metals Park, Ohio, 1974.

Metals Handbook, 8th Ed., Vol. 10, "Failure Analysis and Prevention," American Society for Metals, Metals Park, Ohio, 1975.

Metals Handbook, 8th Ed., Vol. 11, "Nondestructive Inspection and Quality Control," American Society for Metals, Metals Park, Ohio, 1976.

Prevention of Structural Failures, Materials/Metalworking Technology Series, No. 5, American Society for Metals, Metals Park, Ohio, 1975.

Colangelo, V. J., and Heiser, F. A., *Analysis of Metallurgical Failures,* John Wiley and Sons, New York, 1974.

REFERENCES

1. Colangelo, V. J., and Heiser, F. A., *Analysis of Metallurgical Failures,* John Wiley and Sons, New York, 1974, p. 147.
2. Rosenthal, D., "Influence of Residual Stress on Fatigue," in *Metal Fatigue,* McGraw-Hill Book Co., New York, 1959, p. 170.
3. Fuchs, H. O., "Techniques of Surface Stressing to Avoid Fatigue," in *Metal Fatigue,* ibid., p. 197.
4. R. Gaydos, "Fatigue in Heavy Machinery," *Met. Eng. Quart.,* Nov. 1963.
5. Hollomon, J. H., and Jaffe, L. D., *Ferrous Metallurgical Design,* John Wiley and Sons, New York, 1947, p. 158.
6. Naumann, F. K., and Spies, F., "Decarburization," in *Case Histories in Failure Analysis,* American Society for Metals, Metals Park, Ohio 44073, 1979, p. 247.
7. Ebert, L. J., Krotine, F. T., and Troiano, A. R., "Why Case Hardened Components Fail," *Metal Progress,* Sept. 1966.
8. Savage, W. F., "What Is a Weld," in *Weld Imperfections,* Addison-Wesley Publishing Co., Menlo Park, Calif., 1968, p. 25.
9. "Arc Welding Processes and Their Application to Low-Carbon Steel," *Metals Handbook,* 8th Ed., Vol. 6, American Society for Metals, Metals Park, Ohio 44073, 1971, pp. 18-21.
10. G. E. Linnert, *Welding Metallurgy — Iron and Steel,* 2nd Ed., American Welding Society, 1949, p. 292.
11. "Weldment: Physical Metallurgy and Failure Phenomena," Proc. 5th Bolton Landing Conf., Aug. 1978, R. J. Christoffel, E. F. Nippes, and H. D. Solomon, Eds., General Electric Co. Tech. Marketing Operation, Schenectady, N.Y., 1979.
12. D. Holt, "The Influence of Weld Defects on Service Performance," in *Source Book in Failure Analysis,* American Society for Metals, Metals Park, Ohio 44073, 1974, p. 120.
13. Allen, D. K., *Metallurgy Theory and Practice,* American Technical Society, Chicago, 1969, p. 593.
14. Ibid., p. 609.
15. Structural Welding Code (AWS) D1.1-75, American Welding Society, 2501 N.W. 7th St., Miami, Fla. 33125.
16. Linnert, G. E., *Welding Metallurgy,* Vol. 2, American Welding Society, New York, 1967, p. 218.
17. Savage, Ref 8, p. 29.
18. *Metals Handbook,* 8th Ed., Vol. 6, American Society for Metals, Metals Park, Ohio 44073, 1971, p. 19.
19. Savage, W. F., Ref 8, p. 30.

20. Allen, Ref 13, p. 610.
21. Irving, R. R., "Can Industry Afford the High Cost of Weld Repair," *Iron Age*, Apr. 7, 1980, p. 49.
22. Naumann, F. K., and Spies, F., "Broken Rear Wheel Suspension," in *Case Histories in Failure Analysis*, American Society for Metals, Metals Park, Ohio 44073, 1979, pp. 25-29.
23. *Case Histories in Failure analysis*, American Society for Metals, Metals Park, Ohio 44073, 1979.
24. *Metals Handbook*, 8th Ed., Vol. 3, "Machining," American Society for Metals, Metals Park, Ohio 44073, 1967.
25. G. F. VanderVoort, "Analyzing Ductile and Brittle Failures," *Met. Eng. Quart.*, Aug. 1976, pp. 50-52.
26. F. B. Stulen and W. C. Schulte, "Fatigue Failure Analysis and Prevention," *Met. Eng. Quart.*, Aug. 1965.
27. Naumann, F. K., and Spies, F., "Fracture of a Bone Drill," in *Case Histories in Failure Analysis*, Ref 23, p. 54.
28. Faust, C. L., "Electropolishing," in *Electroplating Engineering Handbook*, A. K. Graham, Ed., Van Nostrand Reinhold Co., New York, 1971, p. 108.
29. *Reed Thread Rolling Handbook*, Vol. 2, Technical Data, Reed Rolled Thread Die Co., 1978, p. 7.
30. Bakhvalov, G. T., and Turkovskaya, A. V., *Corrosion and Protection of Metals*, Pergamon Press, London, 1965.
31. *Metal Finishing Guidebook Directory*, Annual Edition, Metals and Plastics Publications, Inc., Hackensack, N. J. 07601, 1980.
32. *Source Book in Failure Analysis*, American Society for Metals, Metals Park, Ohio 44073, 1974.
33. Hildebrand, J. F., "Cadmium Embrittlement of High Strength, Low Alloy Steels at Elevated Temperatures," *Materials Protection and Performance*, Sept. 1973, p. 35.

Failure Analysis Methodology As Related to Product Liability

It is not the purpose of this chapter to provide a complete treatise on failure analysis. Such an effort is outside the intended scope of this text, and numerous works[1-7] on the topic of failure analysis currently exist in the literature. Rather, the purpose here is to describe the methodology required for and peculiar to the analysis of a failed product involved in litigation.

The proper application of failure analysis can provide a valuable adjunct to the total engineering input into a product design. The utilization of this technique can point out design errors, materials limitations, and fabrication defects where they exist in the product. From a purely engineering viewpoint, this knowledge can be fed back into the design of the product, improving its reliability and usefulness as a whole.

From a litigation standpoint, the knowledge gained from a failure analysis is an essential ingredient in evaluating the liability of a product. The data collected in such a study should be combined with other less tangible information to create an information data bank against which to evaluate the product. From examination of the product and from other kinds of information available in the data bank, the engineer in concert with the attorney must become familiar with the total product background in order to do a thorough job of evaluation. This background should include the history of the product from its inception as a design concept to its movement through the manufacturing process, including a review of the labeling, packaging, and instruction materials associated with the product. In addition, as will be discussed, the information pool must be supplemented by other less technical information, such as statements, interviews, depositions, etc., which might become available from witnesses. When as much information as possible has been collected, the available facts can be sifted, weighed, and analyzed in an attempt to reconstruct the events which caused the failure.

PROCEDURAL ASPECTS OF EVIDENCE COLLECTION

From the standpoint of failure analysis and product liability, physical evidence consists of any component or tangible material discovered during an examination which would help determine the failure mechanism or the causative factors associated with the failure.

In several ways the process of collecting evidence for product liability litigation differs from that used in ordinary technical analysis. The most significant difference is the requirement of *continuity of evidence* — that is, the maintenance of a continuous record of the possession and condition of the physical evidence. Since several years may elapse before the case comes to trial, it is essential that the condition of the product at the time of failure be established and, if possible, maintained, so that the trier of fact (judge or jury) may evaluate its condition. If changes are made in order to develop data, then these changes must be documented as to who made them, when they were made, and for what purpose. Similarly, the security of the product is of paramount importance. There should be no period of time during which the location and condition of the product is unaccounted for. If such a lapse exists, it can be inferred that changes may have occurred which have altered the product and have impaired its value as evidence. This procedural lapse can have serious consequences. More than one case has been suddenly and dramatically terminated because a critical component was not allowed into evidence. Consequently, you will find woven throughout this chapter a common thread stressing the need for the documentation of the condition of the unit at all stages of the investigation through photographs, drawings, affidavits, depositions, and the maintenance of inventory records and secure storage conditions.

The unit to be examined is very often not the property of the manufacturer, since having been sold it is now the property of the purchaser. This means that the plaintiff's expert will usually have the first opportunity to examine and evaluate the product and the scene, though he is rarely the first member of the investigating team on the scene. Usually, the first person will be an insurance company adjuster or the manufacturer's service representative. It is not only important, but mandatory, that the evidence at this point be preserved and protected from further damage, thereby maintaining a condition representative of the time of failure. If fragments or parts have been found some distance from the site, their location should be noted and the items should be identified and labeled. Because the early procedural aspects are so significant, it is important that a training program be established for those personnel who might be expected to be first on the scene, so that adherence to a valid procedure may be assured.

DOCUMENTARY EVIDENCE

The documentary or written evidence in an investigation is just as important as the physical evidence and in many cases more so. Some of this material is technical, represented by such documents as industrial safety codes and standards, federal regulations and standards, test reports and certifications, engineering and scientific articles and books, design drawings, warranties, and specifications. Pertinent data contained in these records should be abstracted and summarized so that they may be compared with data obtained on the component under investigation. Obviously, these tests and specifications will play a large role in determining whether the product is in conformance with the specification. The original documents should be filed, and if an extensive collection is created, an indexing system should be set up for convenient retrieval.

Information about the operating conditions — for example, temperature, speed, flow rate, etc. — and service history is very important and should also be collected as part of the technical data package. These data on the actual service conditions should then be compared to the design conditions to determine whether any abnormality exists. If such a condition does exist, is it attributable to a design error or is it the result of improper operation or abuse?

There is another body of documentary evidence that the engineer is not usually responsible for collecting, but it comprises a valuable asset during the investigation. This is the testimony of witnesses and interested parties which is available in the form of depositions or interviews. This testimony provides information and a point of view that is not available to the investigator by any other means. This is not to say that all such information is accurate: it is undoubtedly subject to some bias. In fact, the failure may have been caused by some action or lack of action on the part of the witness himself. Ideally, this bias should be disclosed by other testimony or by test data. However, it may not be, and these interviews should not be relied upon solely but should be integrated into the entire package of data developed during the investigation.

CONDUCTING THE FIELD INVESTIGATION

When an engineer or scientist is called upon to perform a failure analysis, he is concerned with determining the origin, the causative factors, and the various responsibilities for the failure. Very often, he is not called in at the time of the occurrence, although this would be the ideal situation. The sequence of the investigation procedure then will vary depending upon where one becomes involved in the action.

Assume for the moment that an engineer is involved in the failure of a piece of machinery operating in the field and is called in shortly after the occurrence. The engineer should approach the failure with an open mind, setting out to collect as much data as possible before attempting to establish the cause.

One of the first steps in this data collection process is to identify and record any information regarding the manufacturer, size, model number, serial number, date of manufacture, etc., of the main component or assembly. Similar information should also be obtained on the pertinent subcomponents. It may well be that the unit in question was not produced by the manufacturer involved in the litigation. This of course would provide the strongest possible defense and it is important that any indication that this is a possibility be gained early in the litigation.

The scene should then be photographed from several angles showing the location of the unit relative to the general layout, including any damage to the unit or deformation visible from a general inspection. Observations relative to the physical condition of the unit or system should be noted and recorded. If necessary, a sketch of the equipment should be made showing the relative location of any damage discerned as well as the general position of the unit. If movement has occurred, the sketch should show the original location as well as any displacement. Photographs and sketches of the site may also be necessary for certain types of failures — for example, those involving explosions, collapses, or multiple failures.

Close-up photographs of fractured components should be taken before the unit is moved, if this is possible, so that the original condition and position of these fractures may be preserved. Fracture surfaces should be padded for protection against mechanical damage.

If the failed unit is portable, it should be appropriately labeled for subsequent identification and removed for laboratory examination. In some cases the unit will be too large to permit its removal in toto, and decisions must be made on which components to remove and how to remove them. These decisions are often the focus of some controversy, as the opposing counsel will often suggest that a significant piece of hardware has been discarded or altered by the removal process. There is no easy answer to this dilemma. The decisions to be made at this point are difficult and often irreversible and will undoubtedly be subjected to criticism. Nevertheless they must be made, and the more thought that goes into the decision the easier it becomes to defend it. The sketches and photographs described above provide supporting evidence of the condition of the unit right after the event.

LABORATORY ANALYSIS

A laboratory analysis made in connection with a product liability investigation is somewhat different from the ordinary failure analysis conducted during routine engineering studies. Although the two have similarities, the product liability analysis differs in that the protocol is more formal, and more detailed records must be kept than for any ordinary in-house investigation.

For units that have been received in a disassembled state, the examination should begin with an inventory of the components. This inventory should include the name of the component, the location of the component in the overall system, and the approximate size of the component. The components should then be tagged or permanently labeled with an identifying number and should be logged into an inventory control system.

A physical description of the components should then be made, keying this description to the identification number previously assigned. The description should include a general discussion of the size and condition of the component together with any unusual

FIG. 9-1. Front (left) and side (right) views of failed component in the as-received condition. (Courtesy of P. DeMeo)

features observable by the naked eye. Here one would describe any identifying serial number or logo on the component.

At this point in the investigation, photographs of the specimens should be taken to show the general form and configuration of the component. As many photographs from as many angles as necessary should be taken to provide a complete description of the component. It has happened that a component has been damaged or lost in transit between one office and another, and the photographic record is all that remains as evidence of its original condition. Consequently, this step is very important. Figure 9-1 shows front and side views of a failed component received as part of an overall examination. These photographs, as well as the description given in Fig. 9-2, would comprise the record of the product as it was received. It should be noted that this example is for illustration only; in an actual suit, it would be the usual practice to take photographs from several angles. Ordinarily black-and-white photographs are adequate. If a specimen displays unusual markings, however, such as corrosion products which can be illustrated better by the use of color, then obviously color photographs should be taken. At this point, you are ready to conduct the technical portion of the analysis.

FIG. 9-2. Laboratory form describing condition of the component pictured in Fig. 9-1.

EVIDENCE INVENTORY FORM

JOB # 81112   CLIENT XYZ Corp.

REC'D FROM   XYZ Corp.   ADDRESS 316 Hamden Ave.
LOCATION   As Above   Elmira, N.Y. 12211

INDIVIDUAL   Richard Roe   ATTN: John Doe
DATE REC'D   3/3/81   FILE # 14- 186121

DESCRIPTION OF ITEM(S) 1 Bumper Hitch Bracket, left side,
5½" H, 3" Wide, 4" Deep, wt= 3.6#, see attached figures.

TOTAL NUMBER OF INDIVIDUALLY STORED PACKAGES   1

DESCRIPTION OF PACKAGE   LOCATION
1. 1 sealed carton 6" X 8" X 4"   A3
2. _____   _____
3. _____   _____
4. _____   _____

ITEM(S) REMOVED FROM INVENTORY   REASON/DATE
DESCRIPTION
              None
_____   _____

_____   _____

ITEM(S) NOT YET INVENTORIED   LOCATION
DESCRIPTION   None

_____   _____

_____   _____

NOTES: Lateral deformation of flanges, approx. 3/8" from
centerline. Cracks observed in corners of flanges.
Deformation and gouging around 5/8" bolt hole. Fracture
remnant of adjacent component still attached to bolt.
Component bears legend Tow Pilot 1297-1.

Plan of Action

The exact procedure to be followed in conducting an examination will depend upon whether the article is a single component or a complete assembly. However, there is one step which applies universally: do nothing right away. Contemplate the article and determine a plan of action. Ideally this should be done before the article arrives, but often this is not possible. However, even if pressed for time, take a portion of the time available to decide what should be done and in what sequence. A general sequence of tests commonly used in evaluating a product failure is presented in Table 9-1. While this table is certainly not all-inclusive, it does serve as a guide in planning the course of action.

TABLE 9-1. Sequence of Tests in Evaluating Failure

I. Determination of prior history
  A. Documentary evidence
    1. Test certificates
    2. Mechanical test data
    3. Pertinent specifications
    4. Correspondence
    5. Interviews
    6. Depositions and interrogatories
  B. Service parameters
    1. Design or intended operating parameters
    2. Actual service conditions
      a. Temperature data (magnitude and range)
      b. Environmental conditions
      c. Service stresses

II. Cleaning

III. Nondestructive tests
  A. Macroscopic examination of fracture surface
    1. Presence of color or texture changes
      a. Temper colors
      b. Oxidation
      c. Corrosion products
    2. Presence of distinguishing surface features
      a. Shear lips
      b. Beach marks
      c. Chevron markings
      d. Gross plasticity
      e. Large voids or exogenous inclusions
      f. Secondary cracks
    3. Direction of propagation
    4. Fracture origin
  B. Detection of surface and subsurface defects
    1. Magnaflux
    2. Dye penetrant
    3. Ultrasonics
    4. X-ray
  C. Hardness measurements
    1. Macroscopic
    2. Microscopic
  D. Chemical analysis
    1. Spectrographic
    2. Spot tests
    3. X-ray diffraction, fluorescence
    4. Electron beam microprobe
  E. Fractography
    1. Transmission electron microscopy
    2. Scanning electron microscopy: X-ray analysis

IV. Destructive tests
  A. Metallographic
    1. Macroscopic: macroetch
    2. Microscopic
      a. Structure
      b. Grain size
      c. Cleanliness
      d. Microhardness
  B. Mechanical tests
    1. Tensile
    2. Impact
    3. Fracture toughness
    4. Special
  C. Corrosion tests
  D. Wet chemical analysis

V. Stress analysis

VI. Storage

If the article is a complete assembly, a functional test, i.e., a test of whether the product is capable of normal operation, is certainly worthy of consideration. A functional test could disclose whether any abnormality in function exists or whether the operation appears normal. One might find with a functional test, for example, that a gas control valve exhibits leakage. After such a discovery, careful thought should be given to whether disassembly should be conducted. Usually this decision will be made by counsel after a consideration of the facts in the case. In some cases — with a system having no prior use, for example — the presence of a malfunction may be considered sufficient evidence of a defect, and counsel may elect not to proceed further at this point. In other cases, in systems with some use, it may be necessary to continue further to determine the exact cause of the malfunction. If disassembly is elected, the steps in the disassembly process should be recorded photographically: that is, the condition of the article at various stages in the disassembly process should be recorded photographically showing the condition and location of the various components within the assembly. The reason that a decision to disassemble an article must be carefully considered can be illustrated by the example above of a gas control valve that leaked during a functional test. The leak could be caused by a piece of debris lodging in the valve seat. The very act of disassembly could dislodge this debris and render the valve functional. On the other hand, if the leak is caused by porosity through the casting walls, the leak is permanent in nature and disassembly will not affect it. Consequently, the possible results of disassembly should be carefully evaluated before any action is undertaken.

Similar care should be used when the article is a single component. If cutting or sectioning is necessary, then a cutting plan such as the one shown in Fig. 9-3 should be developed to show the location from which the specimens have been taken. Even after the tests have been completed the specimens should be identified and stored — for example, a chemical analysis specimen that has been reduced to chips. The authors have learned to their chagrin that an overzealous attorney can assign devious motives to the discarding of even a routine specimen or fragment of the component, even one that has no special significance.

Dimensional Examination

Often, in order to provide data regarding the condition of a component, it is necessary to conduct detailed dimensional evaluations. This is particularly true when the component which has failed is part of a precisely dimensioned machine assembly. A frequently encountered situation where dimensional information is required occurs when one desires to know the degree of straightness of a rotating shaft, either because of bearing damage or because the shaft has failed.

The methodology required in this instance is that consistent with the practice of a good metrology laboratory. The metrology instru-

---

FIG. 9-3. Typical cutting plan showing location of test specimens.

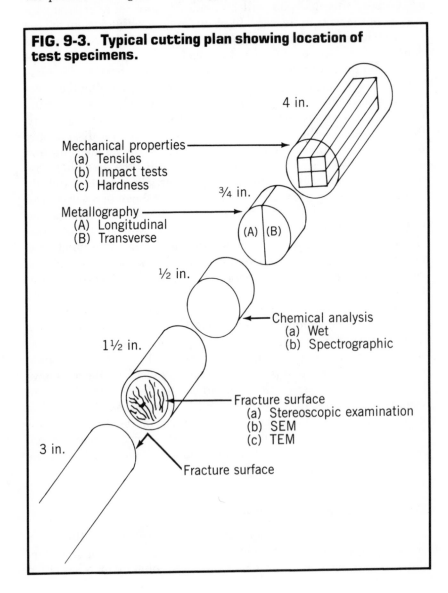

ments, micrometers, calipers, and gauges should be calibrated frequently so that their accuracy cannot be called into question. Similarly, the technique used in the procedure should be recorded and, if necessary, diagrams should be made. The results obtained should be recorded on a drawing of the component, showing the exact location where the dimensional value was obtained.

Nondestructive Testing

Nondestructive testing (NDT) encompasses those test methods that, when applied to a component, do not impair its subsequent utilization. This section is intended to familiarize the reader with various NDT methods, their limitations and the expected results. These methods are particularly important in failure analysis. Structural materials are rarely perfect; usually they contain flaws or defects of varying size, although specifications are customarily applied to limit the severity of these variations. Since a failure often originates from a pre-existing flaw or from one created from the combined action of the environment and the stress state, an important contribution of nondestructive testing is to determine whether the existent flaws or inhomogenities were related to the failure. Another important contribution is to characterize the flaws as to type, size, and location and to determine whether the failed component met the necessary specification requirements. Therefore, standardization of the procedures and the interpretation of the results are extremely important if valid comparisons are to be made.

A number of techniques are used in NDT, each one generally dependent on a different energy system. The techniques range from ordinary macroscopic examination with white light to the complex procedure of neutron radiography. As with testing in other fields, each technique has an area in which it yields optimum performance though it can often be used successfully in marginal situations when the need arises.

Magnetic Particle Inspection. Magnetic particle inspection offers a means for the detection of surface and slightly subsurface discontinuities in ferromagnetic materials. The procedure is not applicable to nonferromagnetic materials, thereby excluding many structural metals, such as austenitic stainless steels, aluminum, magnesium, copper, brass, and titanium, from inspection by this method.

The test is conducted by creating a magnetic field in the test specimen. An indicating medium, either liquid or powder, containing high contrast magnetic particles is applied to the surface, which is then examined for the presence of indications. These indications are caused by the concentration of magnetic particles in the area over the defect. Surface defects appear as sharp indications; subsurface defects yield indications that are broader and less defined. In general, the vagueness increases as the depth below the surface increases.

Defects that can be detected by this method possess a common characteristic — an interface exists between the defects and the bulk material. These include open defects such as cracks, seams, laps, porosity, blow holes, and incomplete fusion of welds.

Liquid Dye Penetrant Inspection. Flaws or defects that intersect the surface, such as cracks, porosity, seams, or laps, can be detected by using liquid penetrants. Subsurface defects are not detectable by this method. Shrinkage, for example, is not usually detectable, unless some prior machining operation has uncovered the defect. Liquid penetrant inspection may be applied to a variety of metals: aluminum, magnesium steels, stainless steels, brasses, and bronzes, as well as plastics, ceramics, and metallo-ceramics. The penetrants used are compounded so that the viscosity is low, and wetting agents are usually added to allow penetration and adsorption within the flaw interior. The penetrants also contain either a dye or a fluorescent material to increase the visibility of the defect.

The techniques require that the penetrant be applied to the surface to be tested and remain there for an appropriate length of time. The material is then flushed off using water or a solvent specific to the system. The surface is dried and a developer is then applied. The function of this developer is to pull the indicating material out of the defect and into the developer coat immediately over it. When dye penetrants are used, the contrast between the defect and background is strikingly apparent under ordinary light conditions. When fluorescent penetrants are used, the parts must be viewed under ultraviolet light to render the flaws visible. Cracks are visible as line indications, and porosity or shrinkage appears as a grouping of point indications. The width of the flaw has a direct bearing on the ease of visibility, the tighter cracks being less prone to discovery.

Radiographic Examination. The basis of radiographic inspection is that X-rays can pass through materials that are optically opaque. The absorption of the incident X-ray by the material is a function of the thickness and nature of the material and the intensity of the incident radiation. The transmitted radiation is dependent on the thickness and nature of the material through which it passes;

therefore, it can be used for the detection of internal flaws that may also have external manifestations, such as a shrinkage cavity that has a surface connection.

A radiograph is the permanent record obtained when a sensitized film is exposed to X-rays passing through a test specimen. The degree of exposure is proportional to the intensity of the transmitted radiation at each point. Flaws such as cracks and voids, being hollow, will therefore show up as darker areas, whereas refractory inclusions will appear as lighter areas, since they tend to absorb a greater amount of radiation. The absorption by a material depends primarily on its atomic number, density, and to a lesser extent its thickness. The ability of radiographic procedures to detect a flaw depends on the density of the flaw, its size, geometry, and orientation. As a check on the adequacy of the radiographic technique, a standard test specimen or penetrameter is placed on the source side of the subject. This test device is usually of a simple geometric configuration—for example, a step block, a series of drilled holes, or a series of wires, made of a material radiographically similar to the subject.

Various types of defects can be identified. Perhaps the major use of radiographic inspection is in the examination of castings, undoubtedly because they are very prone to internal defects. In castings, the procedure can be used to identify and assess such defects as cracks, cold shuts, shrinkage porosity, misruns, gas holes, inclusions, severe segregation, hot tears, and core shifts. Welding and forging defects that involve void space and inclusions obviously can also be inspected by radiography. Specific radiographic examples can be found in various handbooks where casting, forging, and welding defects are discussed in detail (for example, Nondestructive Testing Handbook, Volume I, ASTM Annual Standards, and the ASME Boiler and Pressure Vessel Codes, Section V). Radiography or fluoroscopy can also be utilized in the evaluation of total assemblies to determine whether an assembly is correct, and this information can be obtained without disassembly.

Eddy Current Inspection. The eddy current technique and its variations are very useful in detecting defects at or near the surface of the component being tested. The technique is not confined to ferrous materials but can be used on any metal system. The method is based on the induction of eddy currents in the test specimen when a test coil bearing alternating current is placed on the metal surface. The magnitude of the response is dependent on the frequency and amplitude of the current, the magnetic permeability and electrical conductivity of the material being tested, and the size, shape, and orientation of the defects.

Since the procedure is sensitive to many variables, standardization of the test procedure is a prerequisite for any change in material or geometry. When large numbers of components similar in material or geometry must be tested, however, the procedure is ideal. Precisely because it is sensitive, this method can be utilized to determine the presence of a number of defective situations when comparative standards are available. Surface and subsurface cracks are easily detectable, but the procedure has also been used to determine the thickness of various other types of surface discontinuities.

Ultrasonic Testing. Ultrasonic testing involves the generation of ultrasonic waves, usually by means of a piezoelectric crystal coupled to the surface with a liquid medium, and subsequent measurement of the wave characteristics. There are three main types of ultrasonic waves: longitudinal, shear (transverse), and surface. With each wave type, a number of procedures have been devised, each generally directed at a specific goal. For example, in highly attenuating materials, the through-transmission method is quite useful. In this technique, a receiving transducer is placed in a position to receive the wave emitted from a transmitting transducer. In defect-free materials, this energy is essentially undiminished in magnitude. When a flaw exists, the wave is changed as a result of acoustical impedance, and the signal received is diminished proportionately. If the defect is sufficiently large, the signal will be totally extinguished. The obvious disadvantages of this procedure are that accessibility to two surfaces is required and that the procedure gives no information about the location of the defect. One advantage, however, is that it can be used in energy-limited situations, such as with large castings or with large grain size material.

Thickness measurements can also be determined from one side of a component by using the resonance method of ultrasonic inspection. Several techniques are in general use. Of these, the most widely known is the basic technique — pulse echo. In this technique, the transducer emits a high frequency pulse, then stops to receive the echo of that pulse either from the opposite side of the specimen or from an intervening defect. These signals are usually displayed on a cathode ray tube after suitable conversion and amplification. The travel time of the reflected wave is accurately determined and appears as a pulse to the right of the initial pulse. From its position, the location of a defect within the specimen can be established. In addition, an approximate concept of the size, shape, and orientation of the defect can be determined by checking the specimen from another location or surface.

In general, the higher the frequency is, the greater the sensitivity. However, at the higher frequencies, the sensitivity can be so great that the wave is blocked by such things as grain boundaries, inclusions, and other metallurgical variables. Therefore one must choose between maximum sensitivity and penetration. The pulse echo technique has general applications. Changes in resonance are caused by variations in the wall thickness. Therefore the technique can be used to detect variations in the walls of pressure vessels, tubing, and thin-wall piping caused by process variables or by service conditions — pitting or general corrosion, for example.

The methods just described can be used to determine the presence of internal defects in forgings, castings, welds, and semi-finished products such as bars, rods, and tubing. An excellent presentation on the topic of flaw detection in metal structures has been made by Krautkramer.[8] The test is usually conducted through the smallest dimension of the component — that is, transverse to major axis. In an investigation of product liability, the aim is the detection of flaws or cracks that may have contributed to the failure. The types of defects that can be determined using ultrasonics are those in which a discontinuity or interface exists, such as cracks caused by forging, quenching, fatigue, or hydrogen flaking. Nonmetallic inclusions, those resulting from deoxidation and other chemical reactions, and those caused by the entrapment of foreign materials can be detected using ultrasonics providing these inclusions are sufficiently large relative to the section size being tested.

The primary candidates for detection in castings are internal defects such as shrinkage, blow holes, and porosity. The complexity and finish of the surface, however, can limit the use of ultrasonics for the inspection of castings. Weld defects such as slag inclusions, incomplete penetration, cracking, and porosity can be detected quite readily in most welded structures. It has even become common practice to use ultrasonic testing in field inspection of welds made on certain construction projects, such as bridges.

Cleaning

It would normally be considered standard procedure to clean the fracture surfaces prior to examination in order to reveal surface texture and details. However, even cleaning should not be the result of a decision made in haste. The presence or absence of deposits on the surface may yield significant data regarding the fracture cause and should be noted. If deposits or debris is present, a sample should be collected and labeled for subsequent analysis, when required. Vandervoort[9], for example, describes a failure analysis in which the fracture was covered with paint for a portion of its depth, thereby giving information regarding the age of the crack and its relative rate of growth.

These authors have had a similar experience in which a pipe wrench exhibited a significant shrinkage cavity on the fracture surface. The shrinkage cavity was partially infiltrated with red paint of the same type as that found on the exterior of the wrench, proving that the cavity had an external manifestation and raising the question of whether it should have been detected during scheduled visual inspections.

When cleaning is ultimately decided upon, the procedure should be as gentle as possible, thereby creating no new damage and minimizing the introduction of artifacts on the fracture surface. The cleaning procedure should use a noncaustic cleaning solution or solvent applied by means of a soft plastic or natural fiber brush to loosen and remove any adhering debris. This may be followed by immersion in an ultrasonic bath. If the fracture surface is contaminated with a rust or other corrosion film, it may often be cleaned by the application and stripping of cellulose acetate tape which has been softened by solvent. Repeated applications may be necessary before the specimen is ready for macroscopic examination.

Macroscopic Examination

Macroscopic examination, including the utilization of the stereoscopic microscope, is an extremely useful technique in failure analysis. It is extremely versatile and can be conducted in the field, in the laboratory, or in the courtroom. If a microscope is not available, then a hand lens or even a camera lens can be pressed into service. This procedure is capable of yielding a considerable amount of information about the fracture surface including the following:

> Presence or absence of plasticity (deformation)
> Origin of fracture, single or multiple
> Presence of intergranular fracture
> Presence of secondary cracking
> Surface contamination or corrosion products
> Mechanism of failure

Direction of fracture propagation
Nature and location of stress raisers
Defects on or adjacent to the fracture surface
Estimate of the relative stress level (fatigue)

Assessing the significance of the features seen during macroscopic examination requires the skills of a trained observer. Numerous studies[10-17] have been published describing the kinds of features which may be observed and the mechanisms which may be ascribed to these observations. Certainly anyone conducting a failure analysis should be thoroughly familiar with these findings. Successful failure analysis requires that as many of these studies be acquired as possible, as each may provide an insight into some aspect of failure analysis which has not previously been considered. The point to be emphasized here is not the routine technical considerations involved in failure analysis, which may be learned from a thorough review of the literature, but rather the interpretation of the observations and their relationship to the failure as a whole.

Let us present the following example to illustrate this point. The authors had occasion to determine the cause of failure of a rear wheel hub on a piece of earth-moving equipment. Examination of the fracture surface clearly revealed that the fracture was a fatigue failure caused by bending stresses. The examination also disclosed that the designers had provided a very generous radius at the point where the fracture originated. Furthermore, this area had been shot peened to induce a residual compressive stress state. The creation of a fatigue crack under these circumstances was obviously unusual. The initial examination of the fracture surface had shown that a separate crack originated on each side of the shaft and the progression was unequal. On this basis, a detailed dimensional check was conducted and it was found that the flange surface was out of tolerance with respect to perpendicularity relative to the axis of the shaft. As a result of these findings, an in-depth study of the past history of the particular piece of earth-moving equipment was conducted. It was found that the vehicle had been involved in an off-the-road rear-end collision in which no significant damage was reported.

The wheel hub had apparently been distorted in this prior accident, creating a condition in which each wheel cycle produced excessive bending loads on the hub radius. The significance of the finding is that there was no design or materials defect responsible for the fatigue crack. Since the failure was the result of a prior accident, liability from the product was not an issue.

Microscopic Examination

Microscopic examination can be classified into two types, metallographic and fractographic. Metallographic examination consists of taking a representative section of the component, putting it in a plastic mount, grinding and polishing the surface to a mirror-like finish, then usually etching the surface with a chemical reagent and examining it with a metallographic microscope at magnifications typically from 25× to 1500×.

The preparation and examination of such metallographic specimens is a science in itself. There are a number of fine textbooks[18-20] dealing with the topic. As with the mechanical tests, the preparation of the metallographic specimens should be done in accordance with recognized standards such as ASTM Standard E3-62, "Preparation of Metallographic Specimens." Adherence to these standard methods will prevent the creation of artifacts that might obscure significant details and lead to false conclusions.

This type of examination reveals the microstructure of the material and can be used to determine such things as the kinds and quantities of phases present, the presence or absence of phases at the grain boundaries, grain size, porosity or shrinkage, types and volume of inclusions, thickness of platings and coatings, and case depths. The information gleaned from an examination of the microstructural details can then be compared to a normal structure, either on the basis of past experience or by comparison with representative structures available in such compendiums as the "Atlas of Microstructures of Industrial Alloys."[21]

Occasionally, polished metallographic specimens are examined in the electron beam microprobe apparatus to chemically identify second phase particles (such as nonmetallic inclusions), grain boundary films, and surface layers. This technique allows both a qualitative and quantitative assessment of metallurgical features and defects which may have contributed to a product failure.

A fractographic examination is, as the name implies, an examination of the fracture surface to determine the presence of significant features. To some extent, this has been discussed under the section Macroscopic Examination, but fractography extends beyond the macroscopic range to the microscopic.

Microscopic studies of fracture surfaces are usually conducted using fractographic techniques developed for the transmission electron microscope (TEM) or the scanning electron microscope

(SEM). *Metals Handbook,* Volume 9, "Fractography and Atlas of Fractographs,"[22] describes in detail the application and interpretation of both TEM and SEM techniques.

Because of several desirable characteristics, the SEM has largely supplanted the TEM for failure analysis investigations in recent years. The SEM makes possible direct examination of fracture surface rather than a replica, with little specimen preparation, high resolution of detail, and good depth of field. In addition, the SEM can be fitted with accessories which permit the chemical analysis of microscopic areas on the fracture surface.* This feature alone can provide an indispensable adjunct in a failure analysis. Figure 9-4 illustrates the capability of the SEM. A large void on a fracture surface was examined and found to be the result of interdendritic shrinkage as evidenced by the prominent display of dendrites (Fig. 9-4, top). A continuous interdendritic network, rich in phosphorus and sulfur, was also present (Fig. 9-4, bottom). These photographs also serve to show the high depth of field attainable with SEM fractography.

One disadvantage of the SEM is that the size of the specimen which can be examined is limited by the size of the chamber in the microscope. This means that larger components must be sectioned or replicated before an examination, and as we have previously stated, sectioning of a component requires careful thought.

Another disadvantage of the SEM is that it is an expensive piece of equipment. However, there are increasing numbers of service organizations available in various parts of the country offering time on the machine to be purchased at nominal charges. Since the supervising engineer can accompany the part to be examined, no loss in continuity need occur.

Russo[23] has written an excellent paper describing the usefulness of the SEM in failure analysis. Also presented are a number of case histories showing the exact role played by SEM microscopy in the failure analysis program. There are a number of situations, as in examining the fracture surface of a hypodermic needle or a fractured pacemaker lead, where SEM fractographic techniques literally have no substitute. This is not to say that all failure analyses require such sophisticated equipment. Not so! In most cases, very competent work can be conducted with conventional tools.

Nor should the inference be drawn that the TEM is of no value in fractographic examination. The TEM provides better resolution of the

---

*One such device is known as the EDXA, Energy Dispersive X-ray Analyzer.

FIG 9-4. SEM fractographs of an interdendritic shrinkage cavity at 125× (top) and 400× (bottom), showing sulfur- and phosphorus-rich interdendritic phase.

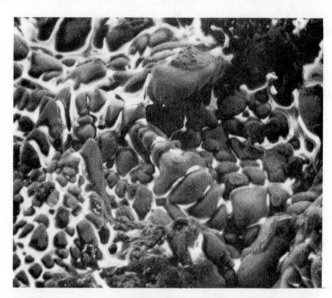

fine details of the fracture surface at very high magnification, such as minute fatigue striations and tear ridges resulting from hydrogen embrittlement. A discussion of the advantages and disadvantages of SEM versus TEM is presented in the ASM *Metals Handbook*.[24] A comparison of four different fracture modes as they appear using SEM and TEM is shown in Fig. 9-5. The difference in depth of field between the two is dramatic. However, the comparison illustrates that either technique can serve quite well in analyzing fracture modes.

One technique commonly used with TEM and also occasionally with SEM fractographic examination is the preparation and examination of a replica of the fracture surface. This technique permits the examination of a fracture surface which otherwise could not be examined because of its massive size or because of restrictions imposed by litigation. Preparation of the replica is a two-stage process. A solvent-moistened plastic film is applied to the fracture surface, allowed to harden, and then removed from the specimen. The plastic replica is placed in a vacuum evaporator for shadowing with an electron opaque material and for formation of a carbon second-stage replica. Separation of the carbon replica from the first-stage plastic replica is accomplished by dissolving the plastic. The replica is now placed on a copper grid for viewing. This simplified description serves to illustrate the manner in which a replica is produced to create a three-dimensional effect as shown in Fig. 9-6. For a complete description of the process, the reader is referred to the ASM *Metals Handbook*, Volume 9,[25] and the U.S. Air Force Fractograph Handbook.[26]

Mechanical Testing

During the course of analyzing the cause of a product failure, it may be necessary to use a number of the conventional tests such as hardness, fatigue, tensile, impact, or fracture toughness tests to characterize the properties of the material.

In conducting these tests, it is extremely important to adhere to recognized standards and procedures such as those prescribed by ASTM.[27] Failure to adhere to these procedures could mean that the results obtained might be declared invalid and rejected as evidence, a determination which could undoubtedly have serious consequences on the outcome of any forthcoming litigation. As regards the interpretation of the results obtained, textbooks such as Dieter's[28] provide a useful reference on the mechanical behavior of metals and test

FIG. 9-5. SEM (left) compared with TEM (right) fractographs of four different fracture modes.

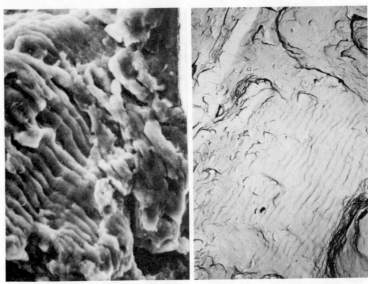

(a) Fatigue fracture, showing fatigue striations; 2500×

(b) Brittle fracture, showing cleavage; 2500×

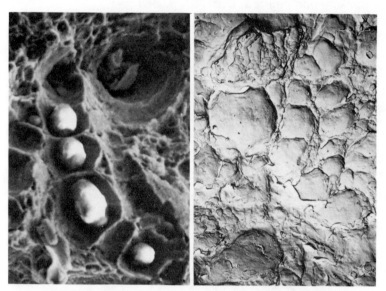

(c) Ductile fracture, showing dimpling; 2500×

(d) Intergranular fracture, showing faceting and secondary cracking; 3500×

FIG. 9-6. Shadowing process for a two-stage plastic-carbon replica. (Ref 22, p. 60)

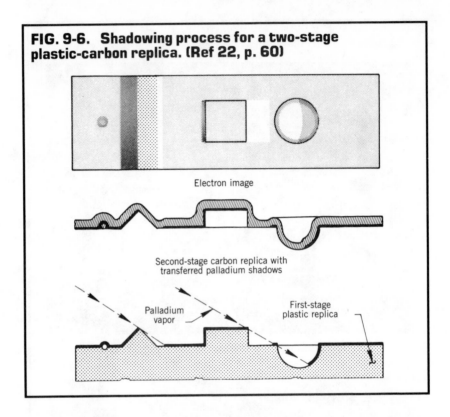

Electron image

Second-stage carbon replica with transferred palladium shadows

Palladium vapor

First-stage plastic replica

procedures, and may suggest additional tests which could further broaden the characterization of the material.

The question of what tests should be included in the schedule will depend upon the product to be examined. Table 9-1 illustrates the tests which would ordinarily be conducted on a metallic component and the general order of testing. Nondestructive tests would normally be conducted first, followed by destructive examination or tests later in the sequence when these tests are deemed appropriate. This sequence is presented for illustration only. The pertinent facts in the case may alter the sequence of the tests and determine whether certain tests should be conducted at all.

When it is decided what tests should be conducted, the location of the test specimens is a matter of immediate concern. Specimens should be taken close enough to the fracture or region of failure to be representative, but not so close as to cause damage to the region in the cutting process. In addition, if gross plastic deformation is involved in the failure, test specimens from this particular region may not be representative of the bulk properties. Unfortunately, there is no perfect solution for this problem. The aggressive attorney will always attempt to imply either that the data are not representative because of the location of the specimen or conversely that the cutting and sectioning significantly altered the remaining pieces so that their value as evidence has been destroyed. One has to rely on one's own judgment as to the potential value of the data generated by the testing, act accordingly, and then be prepared to defend one's actions.

A similar situation exists with fracture surfaces. Though these would not normally be sectioned, there are occasions where sectioning must be considered in order to obtain more information — for example, to permit entry into the scanning electron microscope. Where such sectioning is approved, the most conservative approach would be to reserve one-half of the fracture surface in the untouched condition. The mating half of the fracture surface should be sectioned with a minimum number of cuts. Cuts through the origin should be avoided. The goal should be to generate the maximum amount of information consistent with the minimal amount of damage.

While the above tests are specific for metallurgical components, they also apply to other materials. Hardness tests are applicable to polymers and elastomers, for example. Specialized tests in other materials may also be applicable. In the molding of polycarbonate products, for example, low moisture content is essential for the proper performance of the component. An analysis of the moisture content therefore may well be necessary to determine whether the processing was properly conducted.

Chemical Analysis

The nature of the composition of a material often becomes a question in product liability litigation. Was the material in compliance with the specification? Since this does become an issue, chemical analysis is usually necessary. The following techniques are customarily used:

1. **Wet Chemical Analysis**. This consists of the classical methods of quantitative analytical chemistry in which portions of the component (several grams) are dissolved in solutions, then analyzed in a specified series of steps for each element. The procedures vary with each element but are quite standard. They are usually checked by including National Bureau of Standards control samples with each set of tests.

2. **Spectrographic Analysis**. When a metal is subjected to a high voltage electrical arc, the electrons of the various compositional elements are excited. When these electrons return to their respective ground states, light (visible radiation) is emitted. This emitted light is passed through a diffraction grating, resulting in a spectrum. Examination of the spectrum can reveal how much of an element is present, since each element is represented by a line at a definite wavelength position. The intensity of that line is proportional to the amount of the element present.

3. **Spot Tests**. These are simple analytical tests based on chemical reactions of the compositional elements, with reagents applied to the surface. The presence or absence of the suspected elements is revealed by indicators added to the dissolved spot. Through the application of sequential tests, the composition of an alloy may be verified. Tabulations of various spot tests have been made which facilitate the determination of a number of metals and alloys.[29]

4. **Vacuum Fusion Gas Analysis**. This procedure is used to determine the hydrogen and oxygen content of steels. The method involves melting a small quantity of the metal and extracting the gases. The composition may then be analyzed by one of several methods, including mass spectrometry.

Other techniques such as neutron activation analysis have also recently been employed in special cases, but these are not the tests routinely used to establish composition. They might be used to determine the quantity of metallic contaminants in a coolant system or in a lubrication system, for example.

The techniques previously discussed have largely been those used in analyzing metallic systems. For polymers and elastomers, other instrumental procedures such as infrared analysis and chromatography are used. These are based upon determining the molecular configuration of the organic molecules and comparing these to known spectra and standards.

Stress Analysis

It is quite common during the course of investigating a product failure to conduct an analysis to determine the magnitude of stress existing at various points in the component. This is particularly true if the analysis has not disclosed any gross material defects or if the fractographic evidence indicates the existence of unusually high stresses which were not anticipated.

The procedure here would be to repeat, in essence, those analytical stress calculations which were incorporated into the original design and including any factors which may have been overlooked. The methods used in this procedure would essentially be those discussed in the section on stress analysis in Chapter 4.

STORAGE

Upon completion of the examination and testing, the components should be placed in storage while awaiting further progress of the litigation.

There is some question, even among trained examiners, as to whether the existing fracture surfaces should be coated or treated to prevent additional corrosion during storage. While this would seem like a basic precaution, it could also change the surface and might be construed as an attempt to do so. Here again, the course of action must be that appropriate for the particular case and the particular set of circumstances. Consideration should be given to whether the corrosion inhibitor will interfere with subsequent analysis or alter the composition of surface material. Regardless of whether the surfaces are treated, the fracture surfaces should be protected from mechanical damage by wrapping in plastic or nonabsorbent padding. The storage facilities should be clean and dry and locked, with access only by authorized personnel.

An inventory control system using the assigned laboratory identification number should be developed so that the location of the component is known at all times. While this may appear to be a matter too simple to be worthy of discussion, it should be noted that often a number of transfers take place, from plaintiff's engineer to plaintiff's attorney to defense attorney(s) to defense engineer(s). With each of these transfers, the chance for loss and misplacement exists. Therefore, all transfers should be accompanied by receipts and should be logged into a control system.

REPORTING

When the entire analysis is complete, the usual practice is to report the results obtained in the form of a written report. However, whether such a report is to be written is the option of counsel. Some attorneys prefer that a written report not be issued, operating on the premise that such a report would be discoverable. Others prefer a report, believing that the information is discoverable anyway and the absence of a report prejudicial.

When a written report is issued, the report should include the basic information on the background of the product — i.e., when it was obtained, where it was obtained, from whom it was obtained, and a brief description of the nature of the failure.

The report should describe accurately and simply what was done, and on what components, and the results obtained. The conclusions should be based strictly upon the results obtained and once made should be direct and to the point. They should not be frivolous, speculative, or irrelevant. Do not conclude more than the facts justify.

Above all, the report should be high in readability, that characteristic of a report which renders it easy to understand. Readability is not directly related to subject matter. It is possible to present highly technical information in a highly readable manner, but it does take effort and commitment. Since the purpose of the report is to communicate knowledge to the attorney, little is gained if the language is so technical that only another engineer can comprehend it.

REFERENCES

1. Colangelo, V. J., and Heiser, F. A., *Analysis of Metallurgical Failures*, John Wiley, New York (1974).
2. Bauer, R. D., and Peters, B. F., *How Metals Fail*, Gordon & Breach, New York (1970).
3. Wulpi, D. J., *How Components Fail*, ASM, Metals Park, Ohio (1966).
4. *Metals Handbook*, 8th Edition, Vol. 10, "Failure Analysis and Prevention," ASM, Metals Park, Ohio (1975).
5. *Alloy Steels*, Republic Steel Corp., Cleveland (1968), pp. 301-356.
6. Lipson, C., *Why Machine Parts Fail*, Penton Publishing, Cleveland (1951).
7. *Case Histories in Failure Analysis*, ASM, Metals Park, Ohio (1979).
8. Krautkramer, H., and Krautkramer, J., *Ultrasonic Testing of Materials*, Springer-Verlag, New York (1969).
9. Vandervoort, G. F., *Metallography in Failure Analysis*, Plenum Press (1977), pp. 33-63.
10. Ibid.
11. Colangelo, V. J., and Heiser, F. A., Ref 1.
12. Wulpi, D. J., Ref 3.
13. Bauer, R. D., and Peters, B. F., Ref. 2.
14. Poluskin, E. P., *Defects and Failures of Metals*, Elsevier, Amsterdam (1956).

15. "The Tool Steel Trouble Shooter," Bethlehem Steel Co. (1952).
16. Shand, E. B., *Glass Engineering Handbook*, McGraw-Hill, New York (1958).
17. Andrews, E. H., *Fracture in Polymers*, American Elsevier, New York (1968).
18. Kehl, G. F., *Principles of Laboratory Practice*, McGraw-Hill, New York (1949).
19. McCall, J. L., and Franch, P. M., *Metallography in Failure Analysis*, Plenum Press, New York (1977).
20. *Metals Handbook*, 8th Edition, Vol. 8, "Metallography Structures and Phase Diagrams," ASM, Metals Park, Ohio (1973).
21. *Metals Handbook*, 8th Edition, Vol. 7, "Atlas of Microstructures of Industrial Alloys," ASM, Metals Park, Ohio (1972).
22. *Metals Handbook*, 8th Edition, Vol. 9, "Fractography and Atlas of Fractographs," ASM, Metals Park, Ohio (1974).
23. Russo, M., "Analysis of Fractures Utilizing the SEM," Symposium on Metallography in Failure Analysis, Houston, Texas (July 1977).
24. Ref 22, pp. 63-64.
25. Ref 22, pp. 54-63.
26. Electron Fractography Handbook, Air Force Materials Laboratory, Tech. Report ML-TDR 64-416 (1965).
27. ASTM Standards, Part 10, "Metals — Physical, Mechanical, Corrosion Testing" (1979).
28. Dieter, G., *Mechanical Metallurgy*, McGraw-Hill, New York (1961).
29. Obrzut, J. J., "Tests That Will Help to Identify Metals," *Iron Age*, Nov. 13, 1978, p. 55.

10

Future Directions in Product Liability

The best and cheapest insurance is unquestionably a well-engineered, well-manufactured safe product.[1]

This final chapter will serve to summarize certain aspects of the technical side of product liability. It will also serve as a basis for discussing the prospective directions in product liability with regard to *industry, consumers, engineering,* and *government.* Why are these categories worth discussion? Outside the legal and judicial areas, these are the groups most affected by or involved in the product liability issue. As long as product liability problems persist, we in the technical community must strive to become better acquainted with their various aspects. Such edification is necessary if we are to make a significant contribution to reducing the specter of product liability in society and industry today.

INDUSTRY

There remains little doubt that industry as a whole has been seriously affected by the consequences of product liability, some businesses more severely than others. We previously discussed some details of the impact of product liability on industry in Chapter 1. It is therefore an appropriate time to outline certain directions that industry appears to be headed in.

Costs to Industry

How often we read in headlines or captions that a company is sued for damages in a product liability suit. It is all too frequent a circumstance in this day and age. Furthermore, the cost of such litigation, as we indicated in Chapter 1, is increasing at an alarming rate. These costs include awards, settlements, court fees, lawyers' fees, and investigative fees. In addition, other related costs to manufacturers are significantly increasing. Not only does the producer have to cope with inflation per se, but he must also contend with escalating product liability insurance premiums.

Product Liability Insurance. The problem of product liability insurance has grown rapidly in the last decade. According to the statistics of the Insurance Services Office, average insurance rates for product liability protection increased by 83% from 1974 to 1976. The rates in some industries such as chemical, drugs, sporting goods, machinery, paints, and toys have risen considerably more. Interestingly, these studies reveal that a large number of claims were based on work-related injuries. Furthermore, the three most frequently alleged bases of product liability were:[2]

Design defects. 39%
Manufacturing defects 37%
Failure to warn 21%

Seemingly, we are faced with a "Catch-22" situation. Product liability claims are increasing. The costs of litigation are increasing and, correspondingly, the cost of product liability insurance is rising. In particular, small manufacturers like Sarlo Power Mowers (see Chapter 1) are finding it increasingly hard to afford the protection of product liability insurance. Indeed, many businesses are carrying a very limited amount, or in some cases none at all. From the liability standpoint, this is very risky, because the tendency has been toward large awards or settlements. One or two product liability litigations could potentially *wipe out* a small company.

Even the large companies must be wary of the costs related to product liability suits. Although they may seem better able to absorb the exorbitant costs of litigation, such appearances may generate excessive and false claims. For example, a mobile home fire which took the life of a mother and seriously injured her two children was caused by a propane gas leak.[3] The source of the fire was traced to the furnace. The plaintiff's attorney filed suits against the manufacturer of the mobile home, the furnace manufacturer, and the manufacturer of the controller on the furnace. An expert for the plaintiff discovered positive evidence that the leaking gas was due to a loose inlet fitting on the furnace; thus, the mobile home manufacturer who assembled the system was responsible. The plaintiff's attorney evidently decided not to reveal this evidence because the mobile home manufacturer was a small firm and did not have adequate assets or insurance coverage to satisfy the settlement that the attorney sought. On the other hand, the manufacturer of the furnace controller was a large, nationally known company. Their experts were not able to uncover the same evidence, nor were they able to show that the controller was not defective because it had been damaged by the fire. The conclusion was that the controller manufacturer eventually agreed to an out-of-court settlement of a *few million dollars.*

It is apparent that product liability protection will continue to increase in cost. This is true whether a company chooses to purchase insurance from commercial insurance firms, join a common industrially organized protection fund or "pool," or "self-insure." We note that the National Electric Manufacturing Association has established a product liability program which provides assistance to its members, including sources of expert witnesses, product liability insurance, and defense against product liability suits.[4] If a company is small, Gray[5] suggests that it may benefit from joining an industrial association for the purposes of product liability protection. The following options are then available to the association:

1. Purchase insurance from a selected firm. Each member negotiates his own policy.
2. Mass marketing — joint negotiation by the association as a whole.
3. Industrial group policy — a single policy written for all members.
4. Industrial group policy overriding existing coverage — broader coverage (umbrella-type policy).
5. Set up a mutual or trust fund within the association members (pooling).
6. Industry "captive" insurance firm — serves only the members of an industrial association.
7. Utilize standby bank credit — some leading banks offer standby credit with limits high enough to protect a fund or captive insurance firm.

Recently, the Commerce Department has spoken out on the issue of product liability insurance.[6] For the short term, the department rejected *any* federal programs regarding insurance premium assistance. Instead, they propose a tax set-aside law to permit deduction from taxes of capital. Such deductions would be used to fund a reserve for product liability claims.

For the long term, the Commerce Department recommends eight major areas of federal involvement in the product liability insurance issue. Principally, they suggest a model law that could provide uniformity among states with regard to a statute of limitations (the length of time a manufacturer is responsible for his product), and the relevance of the user's conduct in awarding damages.

Other Costs. In addition to the direct costs of product liability litigation and protection, there are other indirectly related costs to manufacturers. These include the expense of product recalls, rework, and increased quality control.

Product Recalls. Under Section 15(B) of the Consumer Product Safety Act (Appendix 2), manufacturers are required to notify the Consumer Product Safety Commission (CPSC) if one of their products fails to comply with a safety rule or contains a defect which could create a substantial product hazard. During the fiscal year 1978 almost 16 million product units were subject to recall under this section of the law. These products were either repaired or replaced, or refunds were given to the consumer. Undeniably, such activities significantly increase costs to manufacturers. Moreover, manufacturers who do not comply with this section of the law, i.e., fail to notify the CPSC of possible hazards in a timely manner, are subject to civil penalties of up to $500,000 and criminal penalties of up to $50,000 and one-year imprisonment.

The Quality Aspect

Juran[7] defines the *quality function* as "the entire collection of activities through which we achieve fitness for use, no matter where these activities are performed." In other words, he sees quality being produced throughout the entire production cycle, not just associated with *quality control*, which is simply an evaluation of the quality of the component, often applied after the product is completely manufactured. Such a concept is pictorially represented in Fig. 10-1.

We feel strongly that the subject of product quality and product liability are so closely related that any future efforts in avoiding or controlling product liability must necessarily consider the aspect of quality, or the *quality function*. Frankly, the goal of any company

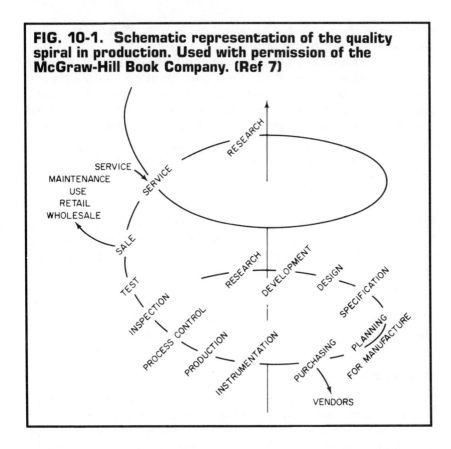

FIG. 10-1. Schematic representation of the quality spiral in production. Used with permission of the McGraw-Hill Book Company. (Ref 7)

should be to avoid product liability claims. When the quality function is properly implemented, the product liability potential of a manufacturer correspondingly decreases. Overall manufacturing costs also decrease, because product conformance increases, thus reducing the amount of recall, rework, and extra testing-inspection that invariably accompanies substandard quality. This outlook is apparently gaining support in certain industries where quality assurance is viewed as a means of keeping out of court.[8]

Our point is that quality can be designed into a component or a system during its conceptual stages of design. This is accomplished, as we discussed in Chapters 4 and 5, by the proper integration of mechanical, geometrical, and materials design procedures. The quality of that component must then be maintained through the processing stages. Sometimes this processing aspect of quality is appropriately referred to as the *quality of conformance*.[9] When such integrity is upheld, the number of nonconforming and thus the number of defective products decreases. The potential for product liability claims against a manufacturer diminishes accordingly.

Recently, a seminar on quality control was held in this country under the sponsorship of the Electronic Industries Association of Japan. The message from Japan was short and unpleasant: "Japan's quality control is better [than ours] and productivity is higher." An article on the seminar reported the Japanese view that:

> The key to quality control is not to increase production line tests and inspection to weed out inferior products. Rather, eliminate potential failure sources at the point of product design (or manufacture), thus precluding the need for extensive inspection and testing.[10]

The report goes on to say that the Japanese see the American view of quality control as one of test, test, test, and then throw out the failures.[11] Their approach, on the other hand, is to eliminate the failures by *doing it right*, and then to eliminate the testing. The ultimate objective is to use quality control to increase productivity. Our suggestion is to use the quality function approach to minimize the potential for defective products reaching the consumer, thereby minimizing the potential for product liability claims.

Management

We wish to draw a parallel between product liability and quality assurance with respect to the management of a company. Usually the management of a company reserves the authority to make policy and decisions regarding the quality level of a product. They are ultimately responsible for defective products, then, because they can effect changes in policy which can prevent product defects from occurring. This same line of reasoning can and should be extended to product liability.

Management of a company can establish policy which reduces the product liability potential of their materials and components. Their authority to do this far outweighs the effectiveness of any other body or organization in reducing product liability.

CONSUMERS

"*Consumers Please Squawk*," the headlines read.[12] It certainly is no secret that consumers are heartily encouraged by the CPSC to complain about products that appear to be defective or substandard.

Even though it is reported that consumers tend to blame themselves for injuries which were not really their fault, the CPSC responded to over 500,000 consumer questions last year. These queries regarding product safety were answered over their toll-free hot line (800-638-8326). In addition, the CPSC responded to some 200,000 letters and product complaints, while distributing over 5 million pamphlets and brochures regarding product safety. Is the American consumer becoming more safety conscious? Unquestionably, he is fast becoming more informed regarding product safety and, correspondingly, product defects.

The goal of making consumers more safety conscious is noble indeed. Everyone would benefit from a reduction in product-related injuries. One unfortunate aspect of this campaign, however, may be the encouragement of perfidious product liability claims. Although we certainly understand and advocate bona fide product liability claims, consumers with weakly supported allegations are now in a more favorable position to initiate product liability litigation against manufacturers. Supported by an unscrupulous attorney, this type of plaintiff can definitely contribute to the growing product liability problem.

Consumer Education

Clearly, the concentrated efforts of the CPSC are having a positive effect on making the consumer aware of product safety. Should such an enormous task be left to this agency and a handful of periodicals? We believe that manufacturers and processors of engineering components or systems are also in a favorable position to inform the public regarding product safety. In addition to producing a defect-free part, educating the consumer in its proper, safe use can also be considered *product liability prevention*.

Educating the consumer can be accomplished in many ways. Advertising campaigns could serve two purposes, one to sell the product, the other to inform the consumer on the safe use of that product. We are not suggesting the use of any kind of disclaimer but rather a safety slogan approach. Safety information bulletins and related literature can accompany the sale of a product along with the other owner information and instructions. Instruction and users manuals should contain safe-use information. Audiovisual aids such as movies and slide presentations can be prepared for products that warrant wider publicity. Such instruction and training can be very effective in industrial situations involving machinery and equipment. In-house or in-plant seminars and briefings for the users of certain equipment are also very effective in educating the user. These techniques and others can bring about a reduction in product-related injuries, thus decreasing product liability claims against a company.

ENGINEERING DIRECTIONS

What will the engineer's role be in product liability in the future? Without a doubt, the engineering community in general will be more deeply involved in the impending issues of product liability. The sooner an engineer can appreciate the intricate legal matters that are associated with product liability litigation and the deleterious consequences of such claims, the more his actions will be responsible ones. The reader may recall that in Chapter 1 we essentially equated product responsibility, that is, the duty and obligation of a manufacturer to produce a safe, reliable component, with product liability.

Engineering Education

On several previous occasions in this text we have alluded to a noticeable lack of formal engineering training regarding product liability in the technological institutions throughout this country. The engineering curricula are necessarily weighted in the subjects of science, mathematics, and engineering, because the student is expected to function competently in these areas upon graduation. However, should not engineering students be at least exposed to the basics of product liability and its application to their prospective profession? Students attending law schools and universities teaching law are exposed to the legal principles of product liability in their studies. As noted earlier, we are currently aware of only two universities that have offered courses in product liability in their engineering departments, Carnegie-Mellon University in Pittsburgh and Ohio State University in Columbus. There may be other technical institutions that offer courses related to this subject, but they are not well publicized.

Correspondingly, we endeavor to interlace the technical aspects of product liability in our class lectures in "Materials Science" and "Failure Analysis." We feel this is beneficial for students who enter the technical job market upon graduation (industrial technologists) as well as those who transfer to engineering universities. Frequently, people working in a technical capacity have an opportunity to affect the product liability attached to engineering materials and products.

Their appreciation of the consequences associated with product liability claims will likely make them more responsible workers and help to improve the product liability position of the company they work for.

Great Britain is at present contemplating a plan which they feel will more effectively utilize their engineering professions.[13] Basically, the proposed plan would offer two types of university programs for engineers, one a practical course oriented towards the majority of students and the other a rigorous five-year course for the most capable students.

They go on to recommend that engineers in both groups be registered, but only the latter group would be "professional engineers," able to approve blueprints prior to production. This program appears to be somewhat similar to the five-year "professional" programs offered by certain technological institutes in the U.S., which are intended primarily as preparation for professional practice.

Many technological universities in this country sponsor short courses and seminars, especially in their off seasons — i.e., summer vacation, midterm breaks, etc. This would be an opportunity to initiate product liability education and test the response of their immediate technical community and available student body. We advocate the initiation of any movement in these directions and feel that the universities that get involved will have a jump on those that procrastinate until more official pressures serve to guide their curricula planning. For instance, it has been reported that the Engineers' Council for Professional Development, the organization which accredits engineering curricula, is internally deliberating the examination of classroom time devoted to ethics and professionalism.[14]

On the Job

Most practicing engineers in industry today are familiar with the term product liability. However, this is typically where the association ends. On the whole, engineers have traditionally ignored the legal aspects of product liability. Such circumstances are not extraordinarily difficult to envision, especially in large corporations where engineers are prevalent and the engineering decision-making process is layered and bureaucratic. Responsibility in these organizations tends to be diffuse, and therefore no one person is at fault when a failure or defective product results in a product liability suit.

Present trends in product liability cases indicate that engineers will be held accountable for design errors and processing defects when they are genuinely responsible for such mistakes that have caused product-related injuries and deaths. This point is reinforced by circumstances where engineers are aware of defective products being shipped or sold to meet quotas.[15,16] Such situations do arise, unfortunately, and the engineer should be prepared to deal with them. Training in both the technical and legal aspects of product liability is necessary to properly educate engineers.

Professional societies for engineers and scientists are aware of the problems facing members who resist production of defective materials and components. So-called "whistle-blowers" are usually frowned upon by management and often suffer the consequences for their integrity. In the same way that certain industrial associations provide assistance to members who are involved in product liability suits, the professional societies should provide counsel and support to members who are faced with questions regarding ethics and faulty products. This support should be stronger as the risk associated with a particular product increases. Furthermore, our professional societies are in a favorable position to sponsor and promote product liability education for their members. Seminars, conferences, courses, etc., can be directed toward the issues of product liability facing the technical community. Special sessions of technical conferences can be devoted to the product liability problems that apply to the technical subjects being presented.

Certification

The efforts of Clarence T. Johnson in 1907, before the Wyoming State Legislature, resulted in the first registration requirements for engineers in the United States. Protection for consumers has extended much further back in history. The Code of Laws of Hammurabi, king of Babylon (approx. 1750 B.C.), provided the following ordinances which protected the citizens from faulty construction:[17]

> If a builder erect a house for a man and do not make its construction firm, and the house which he built collapse and cause the death of the owner of the house, that builder shall be put to death.
>
> If it cause the death of a son of the owner of the house, they shall put to death a son of the builder.

The penalties prescribed in this code were, by today's standards, severe punishment for producing a defective product that resulted in injury or death to the user. Such retribution upon the "producer" surely acted as a deterrent to imprudent design or shoddy workmanship. In more modern times, however, certification of engineers is used to provide protection for users and consumers by requiring professionals to demonstrate their competence before they are allowed to practice engineering. A model definition for "The Practice of Engineering" has been suggested by LaLonde[18] as follows:

> ...any professional service or creative work requiring engineering education, training, and experience and the application of special knowledge of the mathematical, physical, and engineering sciences to such professional services or creative work as consultation, investigation, evaluation, planning, design, and supervision of construction for the purpose of assuring compliance with specifications and design, in connection with any public or private utilities, structures, buildings, machines, equipment, processes, works or projects.

Without a doubt, increased competency can have a beneficial effect on the quality of an engineer's work. This is true whether he is certified or not. Yet, it does seem apparent that qualification of engineers for certain types of jobs and industries will promote compliance and reliability in connection with these tasks. Although a general decrease in product liability potential may be encouraged by increased certification and registration of engineers, the industries and manufacturers involved in relatively high risk products would benefit most. By high risk, we mean products which are frequently involved in product liability suits or products which typically draw large financial claims if they fail. Table 1-1 (Chapter 1) indicates the number of injuries associated with particular consumer products, and Table 10-1, below, shows some products that consumers consider relatively hazardous.[19]

TABLE 10-1. Products Considered Hazardous by Consumers (Adapted from Ref 19)

	January 1978 %	June 1978 %	January 1979 %
Baby products (clothing and toys).........	3	6	12
Foods....................................	9	10	12
Automotive category	0	3	5
Drugs...................................	4	5	5
Packaging	10	6	5
Appliances..............................	1	4	4
Health and beauty aids...................	11	3	3
Soaps and cleansers	4	2	1

Throughout this book we have cited numerous circumstances where product defects and defective conditions have resulted with little or no engineering involvement. Therefore, while certification and registration of engineers may promote competency in certain fields, the major ones being mechanical, civil, and electrical, it does not insure the manufacture of defect-free consumer products. In spite of public pressure about poorly designed products and pressure from within the professional engineering circles, it remains to be seen how significant the effect of increased certification would be in reducing the levels of product liability claims.

Personal Liability

As the trend toward an increase in product liability suits continues (about 177% more suits were filed in U.S. district courts in fiscal year 1978 than in 1974), the aspect of personal liability of engineers will become more prominent. Since the majority of cases continue to involve personal injury due to malfunctioning household appliances, tools, and other manufactured products,[20] engineers as well as other technical groups in these industries are likely to come under closer scrutiny in product liability litigation. In fact, it has been suggested by a plaintiff attorney that "something should be done to permit manufacturers to sue their own employees in cases where the employee has been clearly negligent."[21]

During a recent product liability conference, Fogle[22] presented an excellent discussion of the complexities and risks of personal liability which potentially face engineers. He pointed out that for those engineers involved in *design, manufacture, distribution* or *sales*, and *service*, knowledge of present laws is not sufficient. Therefore, the factors which influence personal liability in product liability litigation are most important and we will briefly review them.

Litigation that involves engineers in product liability from a personal liability standpoint is likely to fall into one of the following categories:

Negligent torts
Intentional torts
Criminal liability

We will not attempt a comprehensive discussion of the details of these categories, which should be left to legal texts, but only mention a few appropriate points. With regard to criminal liability, at least four penal code concepts apply to the personal liability of an engineer: manslaughter, negligent homicide, reckless endangerment, and deception.

Involuntary manslaughter may be committed when, without intent to kill, a person engages in unjustifiably dangerous behavior resulting in death to another. There is however, disagreement among U.S. jurisdictions as to the degree of culpability in these cases. For example, some require recklessness, signifying ignorance on the part of a person that he is creating a peril to the life of another (reckless endangerment); others require only that the accused act in a manner that would make a responsible person aware that his conduct is producing a serious, unjustifiable risk — in other words, a killing by inadvertence amounting to culpable negligence. Ballentine[23] defines a negligent act as "one from which an ordinarily prudent person would foresee such an appreciable risk of harm to others as to cause him not to do the act, or do it in a more careful manner." Such a definition comprises the reasonable-man standard, which can be used to test the actions of a person for negligence. For example, one form of the reasonable-man test which can have engineering implications is generally stated as the omission to do something which a reasonable person, influenced by the considerations which ordinarily control human behavior, would do, or to do something which a reasonable, prudent person would not.

Other crimes of homicide are created by the statutory law of some jurisdictions. For instance, many states have enacted negligent homicide laws that relate to deaths in automobile accidents. Negligent homicide has been broadly defined as follows:[24]

> . . . causing the death of a person without apparent intent to kill, but in doing an unlawful act or performing a lawful act in a careless or negligent manner, the danger of causing death being unapparent.

Can this concept have application to the personal liability of engineers? We are not attempting to answer the legal issue here but simply to apprise engineers of the potential implications. The three penal code concepts just introduced — manslaughter, reckless endangerment, and negligent homicide — are interrelated by the aspect of negligence. Although negligence may not readily be defined with accuracy, we feel that the reasonable-man standards should routinely be considered by those engineers who perform such tasks as design, consultation, and fabrication.

In regard to the concept of deception, Black's Law Dictionary gives the following definition:

> ...intentional misleading by falsehood, spoken or acted. Synony-
> mous with fraud. Knowingly and willingly making a false state-
> ment or representation, express or implied, pertaining to a pres-
> ent or past existing fact.[25]

From the standpoint of an engineer's personal liability, the culpability associated with deceit or deception does not carry as grievous
consequences as the former offenses we have mentioned. Nevertheless, it is an offense that can be related to engineering. Consider, for
example, false statements regarding the safety of a particular design.
Such statements may ultimately become part of the advertising literature and campaign for the product. The liability potential of the
engineer is significantly increased by such a sequence of events.

Overall, the association and application of these criminal
offenses to the personal liability of engineers is left to the opinion of
the reader. We feel that it is sufficient to remind engineers that their
actions must necessarily be competent ones, and the *reasonable-man*
standard is a worthwhile concept to consider literally in all their
professional actions. Although individual circumstances (case by
case) may well dictate actual personal liability, the following terse
quotation[26] is presented as a reminder:

> Any engineer who creates a hazardous product through negligent
> performance of his duties is liable for the consequences.

Historically, the personal liability of an engineer has generally
not surfaced during product liability lawsuits. However, this trend
may suddenly be reversed. It is very important then that engineers be
familiar with the pros and cons of their personal involvement in
product liability litigations. Fogle[27] has summarized the arguments
for and against including the engineer in the product liability suit. He
lists the following factors supporting the view that the engineer
should not be included:

1. There may be more than adequate recovery potential against
   other defendants.
2. The plaintiff may wish to avoid negligence as an issue to
   preclude evidence of his own negligence or to preclude
   proofs by the defendant which show that the defendant made
   exemplary efforts to avoid defects.
3. Statutes of limitation are usually less than for strict liability
   in tort, since the latter commence with the injury.
4. It may not be convenient to have the case tried in a court
   where personal jurisdiction can be obtained over the engineer.
5. The proofs for negligent or intentional torts are more involved than for strict liability.
6. Unless an engineer has obvious wealth, a judgment may

prove worthless. While homeowners insurance will not pay such losses, at the same time it does not encourage the basic suit by its availability.

7. Individuals receive favorable consideration from juries.
8. The employer must indemnify the employee for losses attributable to following employer demands.
9. The identity of the engineer may not be obvious from the documents the plaintiff obtains through discovery.
10. The increase in the number of defendants may preclude a comparative negligence award to plaintiff if the plaintiff is found to be contributarily negligent.

Fogle states that the following factors, on the other hand, tend to favor the inclusion of the engineer in litigation:

A – The engineer with responsibility for a given product is usually easily identified from company records which the plaintiff has a right to obtain.

B – If product legislation and strict liability do not produce safe products, more liability may be assigned to the individuals responsible for the product.

C – It costs very little to add parties to a suit.

D – Other defendants may have no assets.

E – Other defendants may no longer legally exist.

F – Bringing in the engineer may enable the suit to be brought in a court which has a history of favoring liability.

G – Punitive damages for intentional torts may increase an otherwise nominal award.

H – The engineer as a party is a less credible witness for the manufacturer.

I – The employer is entitled to indemnification from the employee, if otherwise innocent of wrongdoing.

There are measures which engineers can take to protect themselves from inordinate exposure to product liability suits. Obviously, performing one's duties in a responsible, competent manner is basic. In today's climate of increased concern about liability it may also be well for an engineer to obtain insurance coverage or set up an agreement with his employer precluding personal liability. Personal liability can also be reduced by *discretionally* signing reports and documents which may eventually be used as evidence in a suit.

Engineers should periodically examine their position from a personal liability viewpoint. The legal and technical aspects of individual liability with respect to a job or a product are far from simple. It may behoove engineers, depending on their own circumstances, to engage professional legal assistance in determining the extent of their personal liability and outlining an individual protection plan, if appropriate.

GOVERNMENTAL ACTION

Federal Action — Consumer Oriented

Based upon recent occurrences, it is not a reasonable expectation that consumer-oriented federal agencies such as the CPSC will decrease their activities regarding product evaluation. Rather, it is much more probable that they will increase their activity and their influence will be felt more widely by manufacturers. Indeed, the 1979 annual report of the CPSC contains this comment by Susan B. King, its chairperson: "I believe that this governmental presence in the marketplace is both necessary and effective. It has raised the general level of safety consciousness both of consumers and industry. Direct CPSC action and voluntary activity by manufacturers have made many products safer than they were ten years ago." This statement appears to verify the generally held belief that the CPSC will continue to exercise increasing influence over the design and manufacture of consumer products.

Earlier in this chapter, in the section on product recalls, we cited statistics regarding the number of product units subject to recall during fiscal year 1978 under the provisions of the Consumer Product Safety Act. In addition, during fiscal year 1979 (October 1, 1978, through September 30, 1979), the thirteen regional CPSC offices monitored 111 recalls of hazardous products that failed to meet safety regulations or contained substantial hazards. As previously noted, firms that fail to notify the CPSC of possible hazards can be subject to fines and jail terms.

How are the number of product-related injuries established? The Consumer Product Safety Commission, through the use of the National Electronic Injury Surveillance System (NEISS), will become increasingly aware of product-related injuries. NEISS is a statistically based injury data collection system. It is made up of a sample of hospitals that are statistically representative of hospitals with emergency departments in the United States and its territories. The system serves to provide national estimates of the number and severity of injuries associated with consumer products and treated in hospital emergency departments. The system also serves as a means of locating injury victims so that further information can be gathered concerning the nature and probable cause of the accident. The data are compiled in various product groups. Table 10-2 is an illustration of the tabulation for one product group, household appliances.[28]

TABLE 10-2. Estimated Injury Statistics From National Electronic Injury Surveillance System (NEISS), Jan. 1 to Dec. 31, 1978

Product Description	Estimate	Mean Weighted Severity	Number of Cases Reported
General household appliances	43,917(a)	—	1,641
Vacuum cleaners	8,367	17	325
Wringer washing machines.................	6,996	70	220
Washing machines, not specified	6,442	26	292
Sewing machines or accessories	5,009	25	203
Portable electric fans.....................	2,912	34	110
Water heaters, not otherwise specified	2,617	71	80
Propane, LP, or butane gas tanks or fittings	2,567	284	65
Clothes dryers, not specified	2,469	28	101
Automatic door openers and closers	1,282	27	50
Gas water heaters.........................	936	102	35
Electric heating pads......................	831	119	33
Water heaters, not specified	752	66	18
Water fountains, cooler or heat units........	544	21	24
Floor buffers or waxers	318	98	12
Automatic doors or door openers	314	56	9
Electric brooms	286	12	2
Incinerators, not otherwise specified	221	66	16
Electric clothes dryers without washers	196	16	8
Washing machines without wringers or other dryers	168	13	13
Mangle irons.............................	161	21	4
Drinking fountains	136	26	6
Automatic garage doors or door openers.....	133	10	2
Washing machines with unheated spin dryers	76	11	2
Rug shampooers...........................	48	13	2
Electric blankets or sheets.................	40	36	3
Washing machines with gas dryers..........	29	17	1
Electric incinerators	25	340	1
Incinerators..............................	17	340	1
Water softeners or conditioners (appliances)...........................	11	10	1
Electric water heaters.....................	9	81	1
Gas clothes dryers without washers	5	12	1

(a) Individual totals do not sum to the general category total.

It is expected that the information gathered from this system will guide the CPSC in setting priorities for selecting types of products for further investigation or action which could ultimately lead to the imposition of product modifications or safety standards.

A similar system of statistical sampling is being established by the National Highway Traffic Safety Administration (NHTSA) with regard to motor vehicle accidents. In 1977, the National Accident Sampling System (NASS) was established to provide valid statistics on motor vehicle accidents, based upon a careful sampling plan.[29] When completed, the NASS system will consist of 75 sampling units based in representative counties within the U.S. These units will collect and process data on accidents occurring within that county. The data base thus collected will provide a basis for assessing the effectiveness of motor vehicle standards and accident prevention measures. It also provides representative accident data for any future research on accident causation. The establishment of these accident data collection systems appears to be a verification of the trend that governmental influence will continue to increase.

Federal Action — Producer Oriented

Various manufacturing groups have emphasized that product liability reforms should occur at the federal level. The Interagency Task Force on Product Liability found in its conclusions that federal regulation of product liability law and insurance could facilitate interstate commerce, avoid duplication and conflicting state efforts, and be more effective in reducing insurance costs. All the activity of the government is not negative; the Commerce Department recommends that the government permit businesses to set aside a portion of their income, tax free, as a reserve against product liability problems and to purchase conventional product liability insurance.

For the longer term, the Commerce Department is recommending the development of a "sound tort liability standard." The main aspects are the creation of uniformity among the states regarding the length of time a manufacturer is liable for his product, the relevance of the product users' conduct with respect to tort awards, and the subject of punitive damages.

Voluntary Standards

A review of the literature published by the CPSC and presentations[30] made by its members leads one to believe that the agency's real interest is to encourage and stimulate industry to develop and adopt voluntary standards. In much of this dialogue, there is the sense of an underlying threat that if industry does not comply by producing effective standards, then mandatory standards will be imposed. The authority for the development of such standards is Section 7 of the Consumer Product Safety Act (Appendix 2).

Whether this threat is real or only implied is a matter of conjecture. Considering the vast number of consumer products which are manufactured, it is doubtful that the CPSC has sufficient funding or manpower to produce mandatory standards. This is undoubtedly why it is promoting voluntary standards and enlisting the cooperation of such recognized standards-making organizations as the American National Standards Institute (ANSI) and the American Society for Testing and Materials (ASTM).

At the same time that the CPSC is encouraging the use and development of voluntary standards, the Federal Trade Commission (FTC) is threatening action that could endanger the entire voluntary system. The proposed rule will attempt to establish the procedures by which product standards would be developed. The rule will require developers of standards to establish written procedures. These procedures would include the requirement that notices be sent to representatives of interested groups, i.e., consumers, producers, and environmental groups, at various stages in the development of a standard. The stages would include the intent to develop the standard, the closing date on the standard, and the date that the standard is to become effective. An excellent summary of the proposed changes is presented by Obrzut in *Iron Age*.[31] This summary states:

> Standards developers would be required to keep careful, detailed records of proceedings and actions, and retain them for at least five years after final action.

> The proposed rule also requires that certain disclosures be made in a standard of the product's attributes as well as serious risks in the product's use when such risks are not apparent.

> Procedures have been established in the proposed rule, the full text of which is printed in the Federal Register of Dec. 7, 1978, to permit complainants or aggrieved parties to challenge standards which they believe to be restrictive, deceptive, unfair or anti-competitive.

> The rule further stipulates that the standards developer may have a duty to act upon receiving a request for action, supported by "substantial evidence," that a standard, proposed standard, interpretation, act or practice, does one or more of the following: (1) Raises prices, excludes products and lacks factual basis; (2) ex-

cludes a product that's equivalent in use to products that meet the standard: (3) excludes a product that's inferior in use to products that don't meet the standard and there's a less restrictive alternative; (4) discriminates against a product that's superior in use to other products that do meet the standard by leading buyers to assume that the product is not superior; (5) misrepresents a product's attributes which affect a buyer's purchasing decisions.

On the other hand, a standards developer need not take action on complaints if there's "substantial evidence on the record as a whole" that none of the above conditions exist. In such a case, the standards developer shall notify the complainant of the decision and the reasons behind it within 60 days, and notify him of the existence of the appeal board.

When a standards developer has a duty to act, appropriate action shall be taken within a reasonable time by: (1) withdrawing the standard by issuing notice within 60 days after receiving the request;(2) developing or revising the standard within a timetable to complete the action, and including the timetable in all notices; (3) issuing within 60 days of the filing of a complaint, a notice listing the details of the planned action.

Any such notice is to state the problems and the alleged harm, and the course of action deemed appropriate. It will also describe action already taken.

Where a standards developer has a duty to act but doesn't, it must withdraw from standards development in the product area and stop distributing standards. It must then issue an appropriate notice to that effect.

The proposed FTC rule also specifies conditions for product certification.

The rule further requires a standards developer to establish an impartial appeal board that would hear disputes over violations of procedures and other routines.

The appeal board is to establish and follow written procedures. These procedures must adhere to certain guidelines stipulated by the FTC proposal.

Proponents of the changes charge that the current procedures are unfair in that they exclude the consumer and small businesses from participation and promote unfair competition. Their most serious charge is that because the standards are consensus standards they do not reflect the current state of attainment but are geared to the level of the lowest producer in the industry group. Advocates of the current system believe that while abuses may occur, they are rare and infrequent.

Understandably, these proposed requirements have produced an outcry from the standards-making organizations, ANSI in particular. The substance of their argument is that the rule will stifle the volunteer effort, which is the basis for the development of the voluntary standards system. There is also concern that FTC domination of the standards-making process will at best be costly and time-consuming and at worst be unworkable.

The future of the voluntary standards is as yet undecided. It remains to be seen whether the FTC will subordinate the voluntary standards system.

State Action — Proposed and Recent Changes

It is quite likely that state legislatures, in response to increasing pressure from manufacturers and insurers, will enact legislation restricting the latitude currently existent in product liability laws. In 1977, Irving [32] stated that 28 states were considering legislation which would modify the existing product liability law. The thrust of these proposed changes is concentrated in several action areas. The primary changes sought are:

1. The imposition of a reasonable statute of limitations which would strive to minimize the "long tail" currently attached to some products — this "long tail" being the inordinately long period that a product is in service and during which the manufacturer may still be held liable
2. The use of state-of-the-art at the time of manufacture as a reasonable defense making subsequent design and manufacturing changes inadmissible as evidence
3. The establishment of more reasonable standards regarding the duty of the manufacturer to warn and instruct
4. The elimination of liability when the product has been altered, modified, or misused after initial sale.

Other proposed changes are an attempt to reduce the amount of damages awarded. Some of these changes deal with the elimination or limitation of punitive damages, the reduction of excessive jury awards, the payment of awards over a period of time rather than in a single payment, and the consideration of other insurance awards in the compensation package. This legislation will probably seek to limit the time exposure of manufacturers for the products they produce. Under the current product liability laws, time relief is minimal; a manufacturer may be held liable for a product produced many years ago even though that product was produced in accordance with the state-of-the-art in existence at the time of its production.

Utah has already passed such legislation. According to the Utah Product Liability Act which became effective in May 1977:

(a) The liability of the manufacturer expires 6 years after the date of the original purchase or 10 years after the date of manufacture.

(b) The manufacturer is relieved of liability, if the product is altered, modified, or misused.

(c) The defect must exist or the danger must be inherent in the product at its time of manufacture or sale in order for liability to exist.

(d) The amount of damages sought in an injury suit cannot be specified.

(e) The existence of unreasonably dangerous defects must be established beyond normal use of a product.

(f) Conformity of a product vis-à-vis design and production to a government standard is a reasonable defense to a product liability suit.

The governmental bodies of New York State have proposed similar legislation.[33] This law would establish a 10-year statute of limitations for actions which seek to recover damages for death, personal injury, or property damage involving product liability. The law also contains several procedural aspects which would serve to limit the recovery to the plaintiff. In addition, the proposed legislation recognizes the state-of-the-art as a reasonable defense. At this time, some of the proposed changes in state law have already been instituted, as evidenced by the legislation passed in Utah. The point is that there will be an increasing effort to minimize the exposure of manufacturers to unreasonable legal and financial burdens in order that the manufacturers remain financially healthy and productive. The premise is that this achieves the greatest good for the greatest number.

EPILOGUE

The consequences of reducing product hazards should be clear. Economic losses can be substantially reduced. More important, the human cost of serious physical injuries, deaths, and the attendant suffering of the injured parties and their families can be lessened. The net result of such action, that is, reducing product defects which cause injuries and economic losses, would necessarily be a reduction in product liability claims. This result is beneficial to all concerned.

Unfortunately, although the engineering and science fields, as well as other facets of the consumer products industry, have made considerable technical progress over the last three or four decades, failures and thus product liability claims are still a problem. A recent viewing of a pre-World War II training film produced by Magnaflux Corporation, dealing with magnetic particle testing, vividly reinforced this point. As we watched this movie, which depicts the formation of certain defects in metals during various production stages, we couldn't help observing that although the film was made 40 years ago, the same defects are still being created in manufacturing processes today. The only real difference is that we can more easily detect them nowadays!

For the reader interested in pursuing additional information regarding the various aspects of product liability, we refer to the comprehensive *Iron Age* article by Robert Irving, which includes an extensive bibliography.[34] In addition, a 32-page bibliography of articles and books on the subject of product liability is available from Dr. Richard Moll, University of Wisconsin, 432 Lake Street, Madison, Wisconsin 53706.

Finally, the question of responsibility is not only a principal theme throughout this book but also an important concern in the issues of product quality and product liability. We feel, therefore, that it is most appropriate to end on this topic. A thought-provoking philosophy regarding the subject of responsibility was expressed by Vice Admiral H. G. Rickover in 1963, testifying at Congressional hearings on the loss of the U.S.S. *Thresher*.* The following quote more than adequately expresses our feelings:

> Responsibility is a unique concept: it can only reside and adhere in a single individual. You may share it with others, but your portion is not diminished. You may delegate it, but it is still with you. You may disclaim it, but you cannot divest yourself of it. Even if you do not recognize it or admit to its presence, you cannot escape it. If responsibility is rightfully yours, no evasion, no ignorance, or passing the blame can shift the burden to someone else. Unless you can point your finger at the man who is responsible when something goes wrong, then you never had anyone really responsible.

---

*Ironically, Vice Admiral Rickover repeated these same remarks at the hearings for the Three Mile Island nuclear incident in 1979.

REFERENCES

1. Powell, G. W., and Mahmoud, S., "An Engineer's Overview of Product Liability," in *Metallography in Failure Analysis*, Plenum Press, New York, 1978, p. 291.
2. Insurance Decisions — Product Liability, INA Corporation (HH—1D17), 1600 Arch St., Phila, Pa. 19101.
3. Powell, G. W., and Mahmoud, S., Ref 1, p. 288.
4. Schreiber, H. A., "The Role of Marketing in Product Liability Exposure Control," Products Liability Exposure Control Symposium, Milwaukee School of Engineering, May 16-17, 1977.
5. Gray, I., "Product Liability — A Management Response," AMACOM, 1975, American Management Association, New York, p. 140.
6. *Iron Age*, Apr. 17, 1978, p. 11.
7. Juran, J. M., Ed., *Quality Control Handbook*, 3rd Ed., McGraw-Hill Book Co., New York, 1951, p. 211.
8. Harvey, R. E., "Quality Assurance Is More Than Just Testing," *Iron Age*, Jan. 14, 1980, p. 22.
9. Harris, D. H., and Chaney, F. B., *Human Factors in Quality Assurance*, John Wiley and Sons, New York, 1969, p. 18.
10. Connally, R., "Japanese Make Quality-Control Pitch," *Electronics*, Apr. 10, 1980, p. 81.
11. Ibid., p. 24.
12. Statler, S. M., "Consumers Please Squawk," *Parade*, Apr. 13, 1980, p. 17.
13. "British Plan to Create Elite Category of Engineers," *Design News*, Sept. 24, 1979, p. 5.
14. Powell, G. W., and Mahmoud, S., Ref 1, p. 293.
15. Ibid., p. 290.
16. Peters, G. A., "The Engineer-Lawyer Interface: Abrasive Conflict or Harmonious Interaction?," Proc. Product Liability Prevention (W), 1977, p. 189.
17. Professional Engineering Examination, Vol. 1, The National Council of Engineering Examiners, 1972, p. X1.
18. LaLonde, W. S., Jr., *Professional Engineers' Examination Questions and Answers*, McGraw-Hill Book Co., New York, 1960, p. 5.
19. Harvey, R. E., Ref 8, p. 23.
20. "Product Liability Cases Continue to Increase," *Iron Age*, May 14, 1979.
21. Irving, R. R., "Our National Product Liability Crisis and Why You Are Part of It," *Iron Age*, Aug. 1, 1977, p. 100.
22. Fogle, F. D., "Personal Liability of Engineers," *Proceedings PLP-77E, Product Liability Prevention Conference*, New Jersey Institute of Technology, Newark, N.J. 07102, pp. 129-136.
23. Ballentine's Law Dictionary, 3rd Ed., W. S. Anderson, Ed., The Lawyer's Co-operative Publishing Co., Rochester, N.Y., p. 840.
24. Ibid., p. 840.
25. Black's Law Dictionary, 5th Ed., 1979, West Publishing Co., St. Paul, Minn., p. 366.
26. Fogle, F. D., Ref 22, p. 133.
27. Ibid., p. 134.

28. U.S. Consumer Product Safety Commission, 1979 Annual Report, Washington, D.C., p.1.
29. National Accident Sampling System, A Status Report, U.S. Dept. of Transportation, December 1978.
30. U.S. Consumer Product Safety Commission, Ref 28, p. 7.
31. Obrzut, J., "Will the FTC Shoot Down Voluntary Standards?," *Iron Age*, Aug. 20, 1979, pp. 59-62.
32. Irving, R. R., Ref 21, pp. 112-113.
33. Proposed bills, N.Y. State Assembly 6219B, New York State Senate 5681 (1977).
34. Irving, R. R., Ref 21, pp. 148-150.

APPENDIX 1:

Outline of Regulations Regarding Safety Acts Under Jurisdiction of Consumer Product Safety Commission*

SUBCHAPTER B – CONSUMER PRODUCT
SAFETY ACT REGULATIONS

PART
1105   Submission of existing standards; offers to develop standards; and the development of standards.
1109   Procedural regulations for oral presentations concerning proposed consumer product safety rules.
1110   Procedures for petitioning for rulemaking under section 10 of the Consumer Product Safety Act.
1115   Substantial product hazard notifications.
1116   Policy and procedures regarding substantial product hazards.
1145   Regulation of products subject to other acts under the Consumer Product Safety Act.
1201   Safety standard for architectural glazing materials.
1202   Safety standard for matchbooks.
1207   Safety standard for swimming pool slides.
1301   Ban of unstable refuse bins.
1302   Extremely flammable contact adhesives.
1303   Lead-containing paint and certain consumer products bearing lead-containing paint.
1304   Ban of consumer patching compounds containing respirable free-form asbestos.
1305   Ban of artificial emberizing materials (ash and embers) containing respirable free-form asbestos.
1401   Self pressurized consumer products containing chlorofluorocarbons; requirements to provide the Commission with performance and technical data; requirements to notify consumers at point of purchase of performance and technical data.

---

*Pertinent details of these regulations may be found in the Code of Federal Regulations, 16CFR1009.3, U.S. Government Printing Office, Washington, D.C.

SUBCHAPTER C – FEDERAL HAZARDOUS SUBSTANCES ACT REGULATIONS

1500   Hazardous substances and articles; administration and enforcement regulations.
1505   Requirements for electrically operated toys or other electrically operated articles intended for use by children.
1507   Fireworks devices.
1508   Requirements for full-size baby cribs.
1509   Requirements for non-full-size baby cribs.
1511   Requirements for pacifiers.
1512   Requirements for bicycles.

SUBCHAPTER D – FLAMMABLE FABRICS ACT REGULATIONS

1602   Statements of policy or interpretation.
1604   Applications for exemption from preemption.
1605   Investigations, inspections and inquiries pursuant to the Flammable Fabrics Act.
1607   Procedures for the development of flammability standards.
1608   General rules and regulations under the Flammable Fabrics Act.
1609   Text of the Flammable Fabrics Act of 1953, as amended in 1954, prior to 1967 amendment and revision.
1610   Standard for the flammability of clothing textiles.
1611   Standard for the flammability of vinyl plastic film.
1615   Standard for the flammability of children's sleepwear: sizes 0 through 6X (FF 3-71).
1616   Standard for the flammability of children's sleepwear: sizes 7 through 14 (FF 5-74).
1630   Standard for the surface flammability of small carpets and rugs (FF 1-70).
1631   Standard for the surface flammability of small carpets and rugs (FF 2-70).
1632   Standard for the flammability of mattresses (and mattress pads) (FF 4-72).

SUBCHAPTER E – POISON PREVENTION PACKAGING ACT OF 1970 REGULATIONS

1700   Poison prevention packaging.
1702   Petitions for exemptions from Poison Prevention Packaging Act requirements; petition procedures and requirements.
1704   Applications for exemption from preemption.

SUBCHAPTER F – REFRIGERATOR SAFETY ACT REGULATIONS

1750   Standard for devices to permit the opening of household refrigerator doors from the inside.

APPENDIX 2:
Consumer Product Safety Act, As Amended*

(Public Law 92-573, enacted October 27, 1972; includes amendments in Pl 94-284, enacted May 11, 1976)

To protect consumers against unreasonable risk of injury from hazardous products, and for other purposes.

Be it enacted by the Senate and House of Representatives of the United States of America in Congress assembled,

SHORT TITLE; TABLE OF CONTENTS

SECTION 1. This Act may be cited as the "Consumer Product Safety Act."

TABLE OF CONTENTS

---

*Amendments in italics.

FINDINGS AND PURPOSES

SEC. 2. (a) The Congress finds that—

(1) an unacceptable number of consumer products which present unreasonable risks of injury are distributed in commerce;

(2) complexities of consumer products and the diverse nature and abilities of consumers using them frequently result in an inability of users to anticipate risks and to safeguard themselves adequately;

(3) the public should be protected against unreasonable risks of injury associated with consumer products;

(4) control by State and local governments of unreasonable risks of injury associated with

consumer products is inadequate and may be burdensome to manufacturers;

(5)   existing Federal authority to protect consumers from exposure to consumer products presenting unreasonable risks of injury is inadequate; and

(6)   regulation of consumer products the distribution or use of which affects interstate or foreign commerce is necessary to carry out this Act.

(b)   The purposes of this Act are—

(1)   to protect the public against unreasonable risks of injury associated with consumer products;

(2)   to assist consumers in evaluating the comparative safety of consumer products;

(3)   to develop uniform safety standards for consumer products and to minimize conflicting State and local regulations; and

(4)   to promote research and investigation into the causes and prevention of product-related deaths, illnesses, and injuries.

DEFINITIONS

SEC. 3. (a) For purposes of this Act:

(1)   The term "consumer product" means any article, or component part thereof, produced or distributed (i) for sale to a consumer for use in or around a permanent or temporary household or residence, a school, in recreation, or otherwise, or (ii) for the personal use, consumption, or enjoyment of a consumer in or around a permanent, or temporary household or residence, a school, in recreation, or otherwise; but such term does not include—

(A)   any article which is not customarily produced or distributed for sale to, or use or consumption by, or enjoyment of a consumer,

(B)   tobacco and tobacco products,

(C)   motor vehicles or motor vehicle equipment (as defined by sections 102(3) and (4) of the National Traffic and Motor Vehicle Safety Act of 1966),

(D)   pesticides (as defined by the Federal Insecticide, Fungicide, and Rodenticide Act),

(E)   any article which, if sold by the manufacturer, producer, or importer, would be subject to the tax imposed by section 4181 of the Internal Revenue Code of 1954 (determined without regard to any exemptions from such tax provided by section 4182 or 4221, or any other provision of such Code), or any component of any such article.

(F)   aircraft, aircraft engines, propellers, or appliances (as defined in section 101 of the Federal Aviation Act of 1958),

(G)   boats which could be subjected to safety regulation under the Federal Boat Safety Act of 1971 (46 U.S.C. 1451 et seq.); vessels, and appurtenances to vessels (other than such boats), which could be subjected to safety regulation under title 52 of the Revised Statutes or other marine safety statutes administered by the department in which the Coast Guard is operating; and equipment (including associated equipment, as defined in section 3(8) of the Federal Boat Safety Act of 1971) to the extent that a risk of injury associated with the use of such equipment on boats or vessels could be eliminated or reduced by actions taken under any statute referred to in this subparagraph,

(H)   drugs, devices, or cosmetics (as such terms are defined in sections 201(g), (h), and (i) of the Federal Food, Drug, and Cosmetic Act), or

(I)   food. The term "food," as used in this subparagraph means all "food," as defined in section 201(f) of the Federal Food, Drug, and Cosmetic Act, including poultry and poultry products (as defined in sections 4(e) and (f) of the Poultry Products Inspection Act), meat, meat food products (as defined in section 1(j) of the Federal Meat Inspection Act), and eggs and egg products (as defined in section 4 of the Egg Products Inspection Act).

Except for the regulation under this Act or the Federal Hazardous Substances Act of fireworks devices or any substance intended for use as a component of any such device, the Commission shall have no authority under the functions transferred pursuant to section 30 of this Act to regulate any product or article described in subparagraph (E) of this paragraph or described, without regard to quantity, in section 845(a)(5) of title 18, United States Code.

See sections 30(d) and 31 of this Act, for other limitations on Commission's authority to regulate certain consumer products.

(2)   The term "consumer product safety rule" means a consumer products safety standard described in section 7(a), or a rule under this Act declaring a consumer product a banned hazardous product.

---

*PL 94-284 included this provision:
The Consumer Product Safety Commission shall make no ruling or order that restricts the manufacture or sale of firearms, firearms ammunition, or components of firearms ammunition, including black powder or gun powder for firearms.

(3)   The term "risk of injury" means a risk of death, personal injury, or serious or frequent illness.

(4)   The term "manufacturer" means any person who manufactures or imports a consumer product.

(5)   The term "distributor" means a person to whom a consumer product is delivered or sold for purposes of distribution in commerce, except that such term does not include a manufacturer or retailer of such product.

(6)   The term "retailer" means a person to whom a consumer product is delivered or sold for purposes of sale or distribution by such person to a consumer.

(7)   (A) The term "private labeler" means an owner of a brand or trademark on the label of a consumer product which bears a private label.

(B)   A consumer product bears a private label if (i) the product (or its container) is labeled with the brand or trademark of a person other than a manufacturer of the product, (ii) the person with whose brand or trademark the product (or container) is labeled has authorized or caused the product to be so labeled, and (iii) the brand or trademark of a manufacturer of such product does not appear on such label.

(8)   The term "manufactured" means to manufacture, produce, or assemble.

(9)   The term "Commission" means the Consumer Product Safety Commission, established by section 4.

(10)   The term "State" means a State, the District of Columbia, the Commonwealth of Puerto Rico, the Virgin Islands, Guam, Wake Island, Midway Island, Kingman Reef, Johnston Island, the Canal Zone, American Samoa, or the Trust Territory of the Pacific Islands.

(11)   The terms "to distribute in commerce" and "distribution in commerce" mean to sell in commerce, to introduce or deliver for introduction into commerce, or to hold for sale or distribution after introduction into commerce.

(12)   The term "commerce" means trade, traffic, commerce, or transportation —

(A)   between a place in a State and any place outside thereof, or

(B)   which affects trade, traffic, commerce, or transportation described in subparagraph (A).

(13)   The terms "import" and "importation" include reimporting a consumer product manufactured or processed, in whole or in part, in the United States.

(14)   The term "United States," when used in the geographic sense, means all of the States (as defined in paragraph (10)).

(b)   A common carrier, contract carrier, or freight forwarder shall not, for purposes of this Act, be deemed to be a manufacturer, distributor, or retailer of a consumer product solely by reason of receiving or transporting a consumer product in the ordinary course of its business as such a carrier or forwarder.

CONSUMER PRODUCT SAFETY COMMISSION

SEC. 4. (a) An independent regulatory commission is hereby established, to be known as the Consumer Product Safety Commission, consisting of five Commissioners who shall be appointed by the President, by and with the advice and consent of the Senate, one of whom shall be designated by the President as Chairman. The Chairman, when so designated, shall act as Chairman until the expiration of his term of office as Commissioner. Any member of the Commission may be removed by the President for neglect of duty or malfeasance in office but for no other cause.

(b)   (1) Except as provided in paragraph (2), (A) the Commissioners first appointed under this section shall be appointed for terms ending three, four, five, six, and seven years, respectively, after the date of enactment of this Act, the term of each to be designated by the President at the time of nomination; and (B) each of their successors shall be appointed for a term of seven years from the date of the expiration of the term for which his predecessor was appointed.

(2)   Any Commissioner appointed to fill a vacancy occurring prior to the expiration of the term for which his precedessor was appointed shall be appointed only for the remainder of such term. A Commissioner may continue to serve after the expiration of his term until his successor has taken office, except that he may not so continue to serve more than one year after the date on which his term would otherwise expire under this subsection.

(c)   Not more than three of the Commissioners shall be affiliated with the same political party. No individual (1) in the employ of, or holding any official relation to, any person engaged in selling or manufacturing consumer products, or (2) owning stock or bonds of substantial value in a person so engaged, or (3) who is in any other manner pecuniarily interested in such a person, or in a substantial supplier of such a person, shall hold the office of Commissioner. A Commissioner may not engage in any other business, vocation, or employment.

(d)   No vacancy in the Commission shall impair the right of the remaining Commissioners to exercise all the powers of Commission, but three members of the Commission shall constitute a quorum for the transaction of business. The Commission shall have an official seal of which judicial notice shall be taken. The Commission shall annually elect a Vice Chairman to act in the absence or disability of the Chairman or in case of a vacancy in the office of the Chairman.

(e)   The Commission shall maintain a principal office and such field offices as it deems necessary and may meet and exercise any of its powers at any other place.

(f)   (1) The Chairman of the Commission shall be the principal executive officer of the Commission, and he shall exercise all of the executive and administrative functions of the Commission, including functions of the Commission with respect to (A) the appointment and supervision of personnel employed under the Commission (other than personnel employed regularly and full time in the immediate offices of commissioners other than the Chairman), (B) the distribution of business among personnel appointed and supervised by the Chairman and among administrative units of the Commission and (C) the use and expenditure of funds.

(2)   In carrying out any of his functions under the provisions of this subsection the Chairman shall be governed by general policies of the Commission and by such regulatory decisions, findings, and determinations as the Commission may by law be authorized to make.

(3)   *Requests or estimates for regular, supplemental, or deficiency appropriations on behalf of the Commission may not be submitted by the Chairman without the prior approval of the Commission.*

(g)   (1) The Chairman, subject to the approval of the Commission, shall appoint an Executive Director, a General Counsel, a Director of Engineering Sciences, a Director of Epidemiology, and a Director of Information. No individual so appointed may receive pay in excess of the annual rate of basic pay in effect for grade GS – 18 of the General Schedule.

(2)   The Chairman, subject to subsection (f)(2), may employ such other officers and employees (including attorneys) as are necessary in the execution of the Commission's functions. No regular officer or employee of the Commission who was at any time during the 12 months preceding the termination of his employment with the Commission compensated at a rate in excess of the annual rate of basic pay in effect for grade GS – 14 of the General Schedule, shall accept employment or compensation from any manufacturer subject to this Act, for a period of 12 months after terminating employment with the Commission.

(3)   *In addition to the number of positions authorized by section 5108(a) of title 5, United States Code, the Chairman, subject to the approval of the Commission, and subject to the standards and procedures prescribed by chapter 51 of title 5, United States Code, may place a total of twelve positions in grades GS – 16, GS – 17, and GS – 18.*

(4)   *The appointment of any officer (other than a Commissioner) or employee of the Commission shall not be subject, directly or indirectly, to review or approval by any officer or entity within the Executive Office of the President.*

(h)   (1) Section 5314 of title 5, United States Code, is amended by adding at the end thereof the following new paragraph:

"(59) Chairman, Consumer Product Safety Commission."

(2)   Section 5315 of such title is amended by adding at the end thereof the following new paragraph:

"(97) Members, Consumer Product Safety Commission (4)."

(i)   *Subsections (a) and (h) of section 2680 of title 28, United States Code, do not prohibit the bringing of a civil action on a claim against the United States which—*

(1)   *is based upon—*

(A)   *misrepresentation or deceit before January 1, 1978, on the part of the Commission or any employee thereof, or*

(B)   *any exercise or performance, or failure to exercise or perform, a discretionary function on the part of the Commission or any employee before January 1, 1978, which exercise, performance, or failure was grossly negligent; and*

(2)   *is not made with respect to any agency action (as defined in section 551(13) of title 5, United States Code).*

In the case of a civil action on a claim based upon the exercise or performance of, or failure to exercise or perform, a discretionary function, no judgment may be entered against the United States unless the court in which such action was brought determines (based upon consideration of all the relevant circumstances, including statutory responsibility of the Commission and the public interest in encouraging rather than inhibiting the exercise of discretion) that such exercise, performance, or failure to exercise or perform was unreasonable.

PRODUCT SAFETY INFORMATION AND RESEARCH

Sec. 5. (a) The Commission shall—

(1)   maintain an Injury Information Clearinghouse to collect, investigate, analyze, and disseminate injury data, and information, relating to the causes and prevention of death, injury, and illness associated with consumer products; and

(2)   conduct such continuing studies and investigations of deaths, injuries, diseases, other health impairments, and economic losses resulting from accidents involving consumer products as it deems necessary.

(b)   The Commission may—

(1)   conduct research, studies, and investigations on the safety of consumer products and on improving the safety of such products;

(2)   test consumer products and develop product safety test methods and testing devices; and

(3)   offer training in product safety investigation and test methods, and assist public and private organizations, administratively and technically, in the development of safety standards and test methods.

(c)   In carrying out its functions under this section, the Commission may make grants or enter into contracts for the conduct of such functions with any person (including a governmental entity).

(d)   Whenever the Federal contribution for any information, research, or development activity authorized by this Act is more than minimal, the Commission shall include in any contract, grant, or other arrangement for such activity, provisions effective to insure that the rights to all information, uses, processes, patents, and other developments resulting from that activity will be made available to the public without charge on a nonexclusive basis. Nothing in this subsection shall be construed to deprive any person of any right which he may have had, prior to entering into any arrangement referred to in this subsection, to any patent, patent application, or invention.

PUBLIC DISCLOSURE OF INFORMATION

SEC. 6 (a) (1) Nothing contained in this Act shall be deemed to require the release of any information described by subsection (b) of section 552, title 5, United States Code, or which is otherwise protected by law from disclosure to the public.

(2)   All information reported to or otherwise obtained by the Commission or its representative under this Act which information contains or relates to a trade secret or other matter referred to in section 1905 of title 18, United States Code, shall be considered confidential and shall not be disclosed, except that such information may be disclosed to other officers or employees concerned with carrying out this Act or when relevant in any proceeding under this Act. Nothing in this Act shall authorize the withholding of information by the Commission or any officer or employee under its control from the duly authorized committees of the Congress.

(b)   (1) Except as provided by paragraph (2) of this subsection, not less than 30 days prior to its public disclosure of any information obtained under this Act, or to be disclosed to the public in connection therewith (unless the Commission finds out that the public health and safety requires a lesser period of notice), the Commission shall, to the extent practicable, notify, and provide a summary of the information to, each manufacturer or private labeler of any consumer product to which such information pertains, if the manner in which such consumer product is to be designated or described in such information will permit the public to ascertain readily the identity of such manufacturer or private labeler, and shall provide such manufacturer or private labeler with a reasonable opportunity to submit contents to the Commission in regard to such information. The Commission shall take reasonable steps to assure, prior to its public disclosure thereof, that information from which the identity of such manufacturer or private labeler may be readily ascertained is accurate, and that such disclosure is fair in the circumstances and reasonably related to effectuating the purposes of this Act. If the Commission finds that, in the administration of this Act, it has made public disclosure of inaccurate or misleading information which reflects adversely upon the safety of any consumer product, or the practices of any manufacturer, private labeler, distributor, or retailer of consumer products, it shall, in a manner similar to that in which such disclosure was made, publish a retraction of such inaccurate or misleading information.

(2) Paragraph (1) (except for the last sentence thereof) shall not apply to the public disclosure of (Å) information about any consumer product with respect to which product the Commission has filed an action under section 12 (relating to imminently hazardous products), or which the Commission has reasonable cause to believe is in violation of section 19 (relating to prohibited acts), or (B) information in the course of or concerning any administrative or judicial proceeding under this Act.

(c) The Commission shall communicate to each manufacturer of a consumer product, insofar as may be practicable, information as to any significant risk of injury associated with such product.

CONSUMER PRODUCT SAFETY STANDARDS

SEC. 7 (a) (1) The Commission may by rule, in accordance with this section and section 9, promulgate consumer product safety standards. A consumer product safety standard shall consist of one or more of any of the following types of requirements:

(A) Requirements as to performance, composition, contents, design, construction, finish, or packaging of a consumer product.

(B) Requirements that a consumer product be marked with or accompanied by clear and adequate warnings or instructions, or requirements respecting the form of warnings or instructions.

Any requirement of such a standard shall be reasonably necessary to prevent or reduce an unreasonable risk of injury associated with such product. The requirements of such a standard (other than requirements relating to labeling, warnings, or instructions) shall, whenever feasible, be expressed in terms of performance requirements.

(2) *No consumer product safety standard promulgated under this section shall require, incorporate, or reference any sampling plan. The preceding sentence shall not apply with respect to any consumer product safety standard or other agency action of the Commission under this Act (A) applicable to a fabric, related material, or product which is subject to a flammability standard or for which a flammability standard or other regulation may be promulgated under the Flammable Fabrics Act, or (B) which is or may be applicable to glass containers.*

(b) A proceeding for the development of a consumer product safety standard under this Act shall be commenced by the publication in the Federal Register of a notice which shall—

(1) identify the product and the nature of the risk of injury associated with the product;

(2) state the Commission's determination that a consumer product safety standard is necessary to eliminate or reduce the risk of injury;

(3) include information with respect to any existing standard known to the Commission which may be relevant to the proceeding; and

(4) include an invitation for any person, including any State or Federal agency (other than the Commission), within 30 days after the date of publication of the notice (A) to submit to the Commission an existing standard as the proposed consumer product safety standard or (B) to offer to develop the proposed consumer product safety standard.

An invitation under paragraph (4)(b) shall specify the period of time in which the offeror of an accepted offer is to develop the proposed standard. The period specified shall be a period ending 150 days after the date the offer is accepted unless the Commission for good cause finds (and includes such finding in the notice) that a different period is appropriate.

(c) If the Commission determines that (1) there exists a standard which has been issued or adopted by any Federal agency or by any other qualified agency, organization, or institution, and (2) such standard if promulgated under this Act, would eliminate or reduce the unreasonable risk of injury associated with the product, then it may, in lieu of accepting an offer pursuant to subsection (d) of this section, publish such standard as a proposed consumer product safety rule.

(d) (1) Except as provided by subsection (c), the Commission shall accept one, and may accept more than one, offer to develop a proposed consumer product safety standard pursuant to the invitation prescribed by subsection (b)(4)(B), if it determines that the offeror is technically competent, is likely to develop an appropriate standard within the period specified in the invitation under subsection (b), and will comply with regulations of the Commission under paragraph (3) of this subsection. The Commission shall publish in the Federal Register the name and address of each person whose offer it accepts, and a summary of the terms of such offer as accepted.

(2) If an offer is accepted under this subsection, the Commission may agree to contribute to the offeror's cost in developing a proposed consumer product safety standard, in any case in

which the Commission determines that such contribution is likely to result in a more satisfactory standard than would be developed without such contribution, and that the offeror is financially responsible. Regulations of the Commission shall set forth the items of cost in which it may participate, and shall exclude any contribution to the acquisition of land or buildings.

"Payments under agreements entered into under this paragraph may be made without regard to section 3648 of the Revised Statutes of the United States (31 U.S.C. 529)."

(3) The Commission shall prescribe regulations governing the development of proposed consumer product safety standards by persons whose offers are accepted under paragraph (1). Such regulations shall include requirements —

(A) that standards recommended for promulgation be suitable for promulgation under this Act, be supported by test data or such other documents or materials as the Commission may reasonably require to be developed, and (where appropriate) contain suitable test methods for measurement of compliance with such standards;

(B) for notice and opportunity by interested persons (including representatives of consumers and consumer organizations) to participate in the development of such standards;

(C) for the maintenance of records, which shall be available to the public, to disclose the course of the development of standards recommended for promulgation, the comments and other information submitted by any person in connection with such development (including dissenting views and comments and information with respect to the need for such recommended standards), and

(D) that the Commission and the Comptroller General of the United States, or any of their duly authorized representatives, have access for the purpose of audit and examination to any books, documents, papers, and records relevant to the development of such recommended standards or to the expenditure of any contribution of the Commission for the development of such standards.

(e) (1) *If the Commission publishes a notice pursuant to subsection (b) to commence a proceeding for the development of a consumer product safety standard for a consumer product and if*

(A) *the Commission does not, within 30 days after the date of publication of such no-*

tice, accept an offer to develop such a standard, or

(B) *the development period (specified in paragraph (3) for such standard ends,*

the Commission may develop a proposed consumer product safety rule respecting such product and publish such proposed rule.

(2) If the Commission accepts an offer to develop a proposed consumer product safety standard, the Commission may not, during the development period (specified in paragraph (3)) for such standard —

(A) publish a proposed rule applicable to the same risk of injury associated with such product, or

(B) develop proposals for such standard or contract with third parties for such development, unless the Commission determines that no offeror whose offer was accepted is making satisfactory progress in the development of such standard.

In any case in which the sole offeror whose offer is accepted under subsection (d)(1) of this section is the manufacturer, distributor, or retailer of a consumer product proposed to be regulated by the consumer product safety standard, the Commission may independently proceed to develop proposals for such standard during the development period.

(3) For purposes of paragraph (2), the development period for any standard is a period (A) beginning on the date on which the Commission first accepts an offer under subsection (d)(1) for the development of a proposed standard, and (B) ending on the earlier of —

(i) the end of the period specified in the notice of proceeding (except that the period specified in the notice may be extended if good cause is shown and the reasons for such extension are published in the Federal Register), or

(ii) the date on which it determines (in accordance with such procedures as it may by rule prescribe) that no offeror whose offer was accepted is able and willing to continue satisfactorily the development of the proposed standard which was the subject of the offer, or

(iii) the date on which an offeror whose offer was accepted submits such a recommended standard to the Commission.

(f) *If the Commission publishes a notice pursuant to subsection (b) to commence a proceeding for the development of a consumer product safety standard and if —*

(1) *no offer to develop such a standard is submitted to, or, if such an offer is submitted to the Commission, no such offer is accepted*

by, the Commission within a period of 60 days from the publication of such notice (or within such longer period as the Commission may prescribe by a notice published in the Federal Register stating good cause therefor), the Commission shall—

    (A) by notice published in the Federal Register terminate the proceeding begun by the subsection (b) notice, or

    (B) develop proposals for a consumer product safety rule for a consumer product identified in the subsection (b) notice and within a period of 150 days (or within such longer period as the Commission may prescribe by a notice published in the Federal Register stating good cause therefor) from the expiration of the 60-day (or longer) period—

       (i) by notice published in the Federal Register terminate the proceeding begun by the subsection (b) notice, or

       (ii) publish a proposed consumer product safety rule; or

    (2) an offer to develop such a standard is submitted to and accepted by the Commission within the 60-day (or longer) period, then not later than 210 days (or such later time as the Commission may prescribe by notice published in the Federal Register stating good cause therefor) after the date of the acceptance of such offer the Commission shall take the action described in clause (i) or (ii) of paragraph (1)(B).

BANNED HAZARDOUS PRODUCTS

SEC. 8. Whenever the Commission finds that—

    (1) a consumer product is being, or will be, distributed in commerce and such consumer product presents an unreasonable risk of injury; and

    (2) no feasible consumer product safety standard under this Act would adequately protect the public from the unreasonable risk of injury associated with such product,
the Commission may propose and, in accordance with section 9, promulgate a rule declaring such product a banned hazardous product.

ADMINISTRATIVE PROCEDURE APPLICABLE TO PROMULGATION OF CONSUMER PRODUCT SAFETY RULES

SEC. 9. (a) (1) Within 60 days after the publication under section 7(c), (e)(1), or (f) or section 8

of a proposed consumer product safety rule respecting a risk of injury associated with a consumer product, the Commission shall—

    (A) promulgate a consumer product safety rule respecting the risk of injury associated with such product if it makes the findings required under subsection (c), or

    (B) withdraw by rule the applicable notice of proceeding if it determines that such rule is not (i) reasonably necessary to eliminate or reduce an unreasonable risk of injury associated with the product, or (ii) in the public interest;
except that the Commission may extend such 60-day period for good cause shown (if it publishes its reasons therefor in the Federal Register).

    (2) Consumer product safety rules which have been proposed under section 7(c), (e)(1), or (f) or section 8 shall be promulgated pursuant to section 553 of title 5, United States Code, except that the Commission shall give interested persons an opportunity for the oral presentation of data, views, or arguments, in addition to an opportunity to make written submissions. A transcript shall be kept of any oral presentation.

    (b) A consumer product safety rule shall express in the rule itself the risk of injury which the standard is designed to eliminate or reduce. In promulgating such a rule the Commission shall consider relevant available product data including the results of research, development, testing, and investigation activities conducted generally and pursuant to this Act.

    In the promulgation of such a rule the Commission shall also consider and take into account the special needs of elderly and handicapped persons to determine the extent to which such persons may be adversely affected by such rule.

    (c) (1) Prior to promulgating a consumer product safety rule, the Commission shall consider, and shall make appropriate findings for inclusion in such rule with respect to—

    (A) the degree and nature of the risk of injury the rule is designed to eliminate or reduce;

    (B) the approximate number of consumer products, or types or classes thereof, subject to such rule;

    (C) the need of the public for the consumer products subject to such rule, and the probable effect of such rule upon the utility, cost, or availability of such products to meet such need; and

    (D) any means of achieving the objective

of the order while minimizing adverse effects on competition or disruption or dislocation of manufacturing and other commercial practices consistent with the public health and safety.

(2) The Commission shall not promulgate a consumer product safety rule unless it finds (and includes such finding in the rule)—

(A) that the rule (including its effective date) is reasonably necessary to eliminate or reduce an unreasonable risk of injury associated with such product;

(B) that the promulgation of the rule is in the public interest; and

(C). in the case of a rule declaring the product a banned hazardous product, that no feasible consumer product safety standard under this Act would adequately protect the public from the unreasonable risk of injury associated with such product.

(d) (1) Each consumer product safety rule shall specify the date such rule is to take effect not exceeding 180 days from the date promulgated, unless the Commission finds, for good cause shown, that a later effective date is in the public interest and publishes its reasons for such findings. The effective date of a consumer product safety standard under this Act shall be set at a date at least 30 days after the date of promulgation unless the Commission for good cause shown determines that an earlier effective date is in the public interest. In no case may the effective date be set at a date which is earlier than the date of promulgation. A consumer product safety standard shall be applicable only to consumer products manufactured after the effective date.

(2) The Commission may by rule prohibit a manufacturer of a consumer product from stockpiling any product to which a consumer product safety rule applies, so as to prevent such manufacturer from circumventing the purpose of such consumer product safety rule. For purposes of this paragraph, the term "stockpiling" means manufacturing or importing a product between the date of promulgation of such consumer product safety rule and its effective date at a rate which is significantly greater (as determined under the rule under this paragraph) than the rate at which such product was produced or imported during a base period (prescribed in the rule under this paragraph) ending before the date of promulgation of the consumer product safety rule.

(e) The Commission may by rule amend or revoke any consumer product safety rule. Such amendment or revocation shall specify the date on which it is to take effect which shall not exceed 180 days from the date the amendment or revocation is published unless the Commission finds for good cause shown that a later effective date is in the public interest and publishes its reasons for such finding. Where an amendment involves a material change in a consumer product safety rule, sections 7 and 8, and subsections (a) through (d) of this section shall apply. In order to revoke a consumer product safety rule, the Commission shall publish a proposal to revoke such rule in the Federal Register, and allow oral and written presentations in accordance with subsection (a)(2) of this section. It may revoke such rule only if it determines that the rule is not reasonably necessary to eliminate or reduce an unreasonable risk of injury associated with the product. Section 11 shall apply to any amendment of a consumer product safety rule which involves a material change and to any revocation of a consumer product safety rule, in the same manner and to the same extent as such section applies to the Commission's action in promulgating such a rule.

COMMISSION RESPONSIBILITY— PETITION FOR CONSUMER PRODUCT SAFETY RULE

SEC. 10. (a) Any interested person, including a consumer or consumer organization, may petition the Commission to commence a proceeding for the issuance, amendment, or revocation of a consumer product safety rule.

(b) Such petition shall be filed in the principal office of the Commission and shall set forth (1) facts which it is claimed establish that a consumer product safety rule or an amendment or revocation thereof is necessary, and (2) a brief description of the substance of the consumer product safety rule or amendment thereof which it is claimed should be issued by the Commission.

(c) The Commission may hold a public hearing or may conduct such investigation or proceeding as it deems appropriate in order to determine whether or not such petition should be granted.

(d) Within 120 days after filing of a petition described in subsection (b), the Commission shall either grant or deny the petition. If the Commission grants such petition, it shall promptly commence an appropriate proceeding under section 7 or 8. If the Commission denies such petition it shall publish in the Federal Register its reasons for such denial.

(e) (1) If the Commission denies a petition made under this section (or if it fails to grant or deny such petition within the 120-day period) the petitioner may commence a civil action in a United States district court to compel the Commission to initiate a proceeding to take the action requested. Any such action shall be filed within 60 days after the Commission's denial of the petition, or (if the Commission fails to grant or deny the petition within 120 days after filing the petition) within 60 days after the expiration of the 120-day period.

(2) If the petitioner can demonstrate to the satisfaction of the court, by a preponderance of evidence in a de novo proceeding before such court, that the consumer product presents an unreasonable risk of injury, and that the failure of the Commission to initiate a rule-making proceeding under section 7 or 8 unreasonably exposes the petitioner or other consumers to a risk of injury presented by the consumer product, the court shall order the Commission to initiate the action requested by the petitioner.

(3) In any action under this subsection, the district court shall have no authority to compel the Commission to take any action other than the initiation of a rule-making proceeding in accordance with section 7 or 8.

(4) *In any action under this subsection the court may in the interest of justice award the costs of suit, including reasonable attorney's fees and reasonable expert witnesses' fees. Attorney's fees may be awarded against the United States (or any agency or official of the United States) without regard to section 2412 of title 28, United States Code, or any other provision of law. For purposes of this paragraph and sections 11(c), 23(a), and 24, a reasonable attorney's fee is a fee (A) which is based upon (i) the actual time expended by the attorney in providing advice and other legal services in connection with representing a person in an action brought under this subsection, and (ii) such reasonable expenses as may be incurred by the attorney in the provision of such services, and (B) which is computed at the rate prevailing for the provision of similar services with respect to actions brought in the court which is awarding such fee.*

(f) The remedies under this section shall be in addition to, and not in lieu of, other remedies provided by law.

(g) Subsection (e) of this section shall apply only with respect to petitions filed more than 3 years after the date of enactment of this Act.

JUDICIAL REVIEW OF CONSUMER PRODUCT SAFETY RULES

SEC. 11. (a) Not later than 60 days after a consumer product safety rule is promulgated by the Commission, any person adversely affected by such rule, or any consumer or consumer organization, may file a petition with the United States court of appeals for the District of Columbia or for the circuit in which such person, consumer, or organization resides or has his principal place of business for judicial review of such rule. Copies of the petition shall be forthwith transmitted by the clerk of the court to the Commission or other officer designated by it for that purpose and to the Attorney General. *The record of the proceedings on which the Commission based its rule shall be filed in the court as provided for in section 2112 of title 28, United States Code.* For purposes of this section, the term "record" means such consumer product safety rule; any notice or proposal published pursuant to section 7, 8, or 9; the transcript required by section 9(a)(2) of any oral presentation; any written submission of interested parties; and any other information which the Commission considers relevant to such rule.

(b) If the petitioner applies to the court for leave to adduce additional data, views, or arguments and shows to the satisfaction of the court that such additional data, views or arguments are material and that there were reasonable grounds for the petitioner's failure to adduce such data, views, or arguments in the proceeding before the Commission, the court may order the Commission to provide additional opportunity for the oral presentation of data, views, or arguments and for written submissions. The Commission may modify its findings, or make new findings by reason of the additional data, views, or arguments so taken and shall file such modified or new findings, and its recommendation, if any, for the modification or setting aside of its original rule, with the return of such additional data, views, or arguments.

(c) Upon the filing of the petition under subsection (a) of this section the court shall have jurisdiction to review the consumer product safety rule in accordance with chapter 7 of title 5, United States Code, and to grant appropriate relief, including interim relief, as provided in such chapter. A court may in the interest of justice include in such relief an award of the costs of suit, including reasonable attorney's fees (determined in accordance with section

10(e)(4)) and reasonable expert witnesses' fees. Attorney's fees may be awarded against the United States (or any agency or official of the United States) without regard to section 2412 of title 28, United States Code, or any other provision of law. The consumer product safety rule shall not be affirmed unless the Commission's findings under section 9(c) are supported by substantial evidence on the record taken as a whole.

(d) The judgment of the court affirming or setting aside, in whole or in part, any consumer product safety rule shall be final, subject to review by the Supreme Court of the United States upon certiorari or certification, as provided in section 1254 of title 28 of the United States Code.

(e) The remedies provided for in this section shall be in addition to and not in lieu of any other remedies provided by law.

IMMINENT HAZARDS

SEC. 12. (a) The Commission may file in a United States district court an action (1) against an imminently hazardous consumer product for seizure of such product under subsection (b)(2), or (2) against any person who is a manufacturer, distributor, or retailer of such product, or (3) against both. Such an action may be filed notwithstanding the existence of a consumer product safety rule applicable to such product, or the pendency of any administrative or judicial proceedings under any other provision of this Act. As used in this section, and hereinafter in this Act, the term "imminently hazardous consumer product" means a consumer product which presents imminent and unreasonable risk of death, serious illness, or severe personal injury.

(b) (1) The district court in which such action is filed shall have jurisdiction to declare such product an imminently hazardous consumer product, and (in the case of an action under subsection (a)(2)) to grant (as ancillary to such declaration or in lieu thereof) such temporary or permanent relief as may be necessary to protect the public from such risk. Such relief may include a mandatory order requiring the notification of such risk to purchasers of such product known to the defendant, public notice, the recall, the repair or the replacement of, or refund for, such product.

(2) In the case of an action under subsection (a)(1), the consumer product may be proceeded against by process of libel for the seizure and condemnation of such product in any United States district court within the jurisdiction of which such consumer product is found. Proceedings and cases instituted under the authority of the preceding sentence shall conform as nearly as possible to proceedings in rem in admiralty.

(c) Where appropriate, concurrently with the filing of such action or as soon thereafter as may be practicable, the Commission shall initiate a proceeding to promulgate a consumer product safety rule applicable to the consumer product with respect to which such action is filed.

(d) (1) Prior to commencing an action under subsection (a), the Commission may consult the Product Safety Advisory Council (established under section 28) with respect to its determination to commence such action, and request the Council's recommendations as to the type of temporary or permanent relief which may be necessary to protect the public.

(2) The Council shall submit its recommendations to the Commission within one week of such request.

(3) Subject to paragraph (2), the Council may conduct such hearing or offer such opportunity for the presentation of views as it may consider necessary or appropriate.

(e) (1) An action under subsection (a)(2) of this section may be brought in the United States district court for the District of Columbia or in any judicial district in which any of the defendants is found, is an inhabitant or transacts business; and process in such an action may be served on a defendant in any other district in which such defendant resides or may be found. Subpoenas requiring attendance of witnesses in such an action may run into any other district. In determining the judicial district in which an action may be brought under this section in instances in which such action may be brought in more than one judicial district, the Commission shall take into account the convenience of the parties.

(2) Whenever proceedings under this section involving substantially similar consumer products are pending in courts in two or more judicial districts, they shall be consolidated for trial by order of any such court upon application reasonably made by any party in interest, upon notice to all other parties in interest.

(f) Notwithstanding any other provision of law, in any action under this section, the Commission may direct attorneys employed by it to appear and represent it.

NEW PRODUCTS

SEC. 13. (a) The Commission may, by rule, prescribe procedures for the purpose of insuring that the manufacturer of any new consumer product furnish notice and a description of such product to the Commission before its distribution in commerce.

(b) For purposes of this section, the term "new consumer product" means a consumer product which incorporates a design, material, or form of energy exchange which (1) has not previously been used substantially in consumer products and (2) as to which there exists a lack of information adequate to determine the safety of such product in use by consumers.

PRODUCT CERTIFICATION AND LABELING

SEC. 14. (a) (1) Every manufacturer of a product which is subject to a consumer product safety standard under this Act and which is distributed in commerce (and the private labeler of such product if it bears a private label) shall issue a certificate which shall certify that such product conforms to all applicable consumer product safety standards, and shall specify any standard which is applicable. Such certificate shall accompany the product or shall otherwise be furnished to any distributor or retailer to whom the product is delivered. Any certificate under this subsection shall be based on a test of each product or upon a reasonable testing program; shall state the name of the manufacturer or private labeler issuing the certificate; and shall include the date and place of manufacture.

(2) In the case of a consumer product for which there is more than one manufacturer or more than one private labeler, the Commission may by rule designate one or more of such manufacturers or one or more of such private labelers (as the case may be) as the persons who shall issue the certificate required by paragraph (1) of this subsection, and may exempt all other manufacturers of such product or all other private labelers of the product (as the case may be) from the requirement under paragraph (1) to issue a certificate with respect to such product.

(b) The Commission may by rule prescribe reasonable testing programs for consumer products which are subject to consumer product safety standards under this Act and for which a certificate is required under subsection (a). Any test or testing program on the basis of which a certificate is issued under subsection (a) may, at the option of the person required to certify the product, be conducted by an independent third party qualified to perform such tests or testing programs.

(c) The Commission may by rule require the use and prescribe the form and content of labels which contain the following information (or that portion of it specified in the rule)—

(1) The date and place of manufacture of any consumer product.

(2) A suitable identification of the manufacturer of the consumer product, unless the product bears a private label in which case it shall identify the private labeler and shall also contain a code mark which will permit the seller of such product to identify the manufacturer thereof to the purchaser upon his request.

(3) In the case of a consumer product subject to a consumer product safety rule, a certification that the product meets all applicable consumer product safety standards and a specification of the standards which are applicable.

Such labels, where practicable, may be required by the Commission to be permanently marked on or affixed to any such consumer product. The Commission may, in appropriate cases, permit information required under paragraphs (1) and (2) of this subsection to be coded.

NOTIFICATION AND REPAIR, REPLACEMENT, OR REFUND

SEC. 15. (a) For purposes of this section, the term "substantial product hazard" means—

(1) a failure to comply with an applicable consumer product safety rule which creates a substantial risk of injury to the public, or

(2) a product defect which (because of the pattern of defect, the number of defective products distributed in commerce, the severity of the risk, or otherwise) creates a substantial risk of injury to the public.

(b) Every manufacturer of a consumer product distributed in commerce, and every distributor and retailer of such product, who obtains information which reasonably supports the conclusion that such product—

(1) fails to comply with an applicable consumer product safety rule; or

(2) contains a defect which could create a substantial product hazard described in subsection (a)(2),

shall immediately inform the Commission of such failure to comply or of such defect, unless such manufacturer, distributor, or retailer has actual knowledge that the Commission has been adequately informed of such defect or failure to comply.

(c) If the Commission determines (after affording interested persons, including consumers and consumer organizations, an opportunity for a hearing in accordance with subsection (f) of this section) that a product distributed in commerce presents a substantial product hazard and that notification is required in order to adequately protect the public from such substantial product hazard, the Commission may order the manufacturer or any distributor or retailer of the product to take any one or more of the following actions:

(1) to give public notice of the defect or failure to comply.

(2) to mail notice to each person who is a manufacturer, distributor, or retailer of such product.

(3) to mail notice to every person to whom the person required to give notice knows such product was delivered or sold.

Any such order shall specify the form and content of any notice required to be given under such order.

(d) If the Commission determines (after affording interested parties, including consumers and consumer organizations, an opportunity for a hearing in accordance with subsection (f)), that a product distributed in commerce presents a substantial product hazard and that action under this subsection is in the public interest, it may order the manufacturer or any distributor or retailer of such product to take whichever of the following actions the person to whom the order is directed elects:

(1) To bring such product into conformity with the requirements of the applicable consumer product safety rule or to repair the defect in such product.

(2) To replace such product with a like or equivalent product which complies with the applicable consumer product safety rule or which does not contain the defect.

(3) To refund the purchase price of such product (less a reasonable allowance for use, if such product has been in the possession of a consumer for one year or more (A) at the time of public notice under subsection (c), or (B) at the time the consumer receives actual notice of the defect or noncompliance, whichever first occurs). An order under this subsection may also require the person to whom it applies to submit a plan, satisfactory to the Commission, for taking action under whichever of the preceding paragraphs of this subsection under which such person has elected to act. The Commission shall

specify in the order the persons to whom refunds must be made if the person to whom the order is directed elects to take the action described in paragraph (3). If an order under this subsection is directed to more than one person, the Commission shall specify which person has the election under this subsection.

An order under this subsection may prohibit the person to whom it applies from manufacturing for sale, offering for sale, distributing in commerce, or importing into the customs territory of the United States (as defined in general headnote 2 to the Tariff Schedules of the United States), or from doing any combination of such actions, the product with respect to which the order was issued.

(e) (1) No charge shall be made to any person other than a manufacturer, distributor, or retailer who avails himself of any remedy provided under an order issued under subsection (d), and the person subject to the order shall reimburse each person (other than a manufacturer, distributor, or retailer) who is entitled to such a remedy for any reasonable and foreseeable expenses incurred by such person in availing himself of such remedy.

(2) An order issued under subsection (c) or (d) with respect to a product may require any person who is a manufacturer, distributor, or retailer of the product to reimburse any other person who is a manufacturer, distributor, or retailer of such product for such other person's expenses in connection with carrying out the order, if the Commission determines such reimbursement to be in the public interest.

(f) An order under subsection (c) or (d) may be issued only after an opportunity for a hearing in accordance with section 554 of title 5, United States Code except that, if the Commission determines that any person who wishes to participate in such hearing is a part of a class of participants who share an identity of interest, the Commission may limit such person's participation in such hearing to participation through a single representative designated by such class (or by the Commission if such class fails to designate such a representative).

(g) (1) If the Commission has initiated a proceeding under this section for the issuance of an order under subsection (d) with respect to a product which the Commission has reason to believe presents a substantial product hazard, the Commission (without regard to section 27(b)(7)) or the Attorney General may, in accordance with section 12(e)(1), apply to a district court of the United States for the issuance *of a preliminary injunction to restrain the distribution in commerce of such product pending the completion of such proceeding. If such a preliminary injunction has been issued, the Commission (or the Attorney General if the preliminary injunction was issued upon an application of the Attorney General) may apply to the issuing court for extensions of such preliminary injunction.*

(2) Any preliminary injunction, and any extension of a preliminary injunction, issued under this subsection with respect to a product shall be in effect for such period as the issuing court prescribes not to exceed a period which extends beyond the thirtieth day from the date of the issuance of the preliminary injunction (or, in the case of a preliminary injunction which has been extended, the date of its extension) or the date of the completion or termination of the proceeding under this section respecting such product, whichever date occurs first.

(3) The amount in controversy requirement of section 1331 of title 28, United States Code does not apply with respect to the jurisdiction of a district court of the United States to issue or extend a preliminary injunction under this subsection.

INSPECTION AND RECORDKEEPING

SEC. 16. (a) For purposes of implementing this Act, or rules or orders prescribed under this Act, officers or employees duly designated by the Commission, upon presenting appropriate credentials and a written notice from the Commission to the owner, operator, or agent in charge, are authorized—

(1) to enter, at reasonable times, (A) any factory, warehouse, or establishment in which consumer products are manufactured or held, in connection with distribution in commerce, or (B) any conveyance being used to transport consumer products in connection with distribution in commerce; and

(2) to inspect, at reasonable times and in a reasonable manner such conveyance or those areas of such factory, warehouse, or establishment where such products are manufactured, held, or transported and which may relate to the safety of such products. Each such inspection shall be commenced and completed with reasonable promptness.

(b) Every person who is a manufacturer, private labeler, or distributor of a consumer product shall establish and maintain such records,

make such reports, and provide such information as the Commission may, by rule, reasonably require for the purposes of implementing this Act, or to determine compliance with rules or orders prescribed under this Act. Upon request of an officer or employee duly designated by the Commission, every such manufacturer, private labeler, or distributor shall permit the inspection of appropriate books, records, and papers relevant to determining whether such manufacturer, private labeler, or distributor has acted or is acting in compliance with this Act and rules under this Act.

IMPORTED PRODUCTS

SEC. 17. (a) Any consumer product offered for importation into the customs territory of the United States (as defined in general headnote 2 to the Tariff Schedules of the United States) shall be refused admission into such customs territory if such product—

(1) fails to comply with an applicable consumer product safety rule;

(2) is not accompanied by a certificate required by section 14, or is not labeled in accordance with regulations under section 14(c);

(3) is or has been determined to be an imminently hazardous consumer product in a proceeding brought under section 12;

(4) has a product defect which constitutes a substantial product hazard (within the meaning of section 15(a)(2)); or

(5) is a product which was manufactured by a person whom the Commission has informed the Secretary of the Treasury is in violation of subsection (g).

(b) The Secretary of the Treasury shall obtain without charge and deliver to the Commission, upon the latter's request, a reasonable number of samples of consumer products being offered for import. Except for those owners or consignees who are or have been afforded an opportunity for a hearing in a proceeding under section 12 with respect to an imminently hazardous product, the owner or consignee of the product shall be afforded an opportunity by the Commission for a hearing in accordance with section 554 of title 5 of the United States Code with respect to the importation of such products into the customs territory of the United States. If it appears from examination of such samples or otherwise that a product must be refused admission under the terms of subsection (a), such product shall be refused admission, unless subsection (c) of this section applies and is complied with.

(c)   If it appears to the Commission that any consumer product which may be refused admission pursuant to subsection (a) of this section can be so modified that it need not (under the terms of paragraphs (1) through (4) of subsection (a)) be refused admission, the Commission may defer final determination as to the admission of such product and, in accordance with such regulations as the Commission and the Secretary of the Treasury shall jointly agree to, permit such product to be delivered from customs custody under bond for the purpose of permitting the owner or consignee an opportunity to so modify such product.

(d)   All actions taken by an owner or consignee to modify such product under subsection (c) shall be subject to the supervision of an officer or employee of the Commission and of the Department of the Treasury. If it appears to the Commission that the product cannot be so modified or that the owner or consignee is not proceeding satisfactorily to modify such product, it shall be refused admission into the customs territory of the United States, and the Commission may direct the Secretary to demand redelivery of the product into customs custody, and to seize the product in accordance with section 22(b) if it is not so redelivered.

(e)   Products refused admission into the customs territory of the United States under this section must be exported, except that upon application, the Secretary of the Treasury may permit the destruction of the product in lieu of exportation. If the owner or consignee does not export the product within a reasonable time, the Department of the Treasury may destroy the product.

(f)   All expenses (including travel, per diem or subsistence, and salaries of officers or employees of the United States) in connection with the destruction provided for in this section (the amount of such expenses to be determined in accordance with regulations of the Secretary of the Treasury) and all expenses in connection with the storage, cartage, or labor with respect to any consumer product refused admission under this section, shall be paid by the owner or consignee and, in default of such payment, shall constitute a lien against any future importations made by such owner or consignee.

(g)   The Commission may, by rule, condition the importation of a consumer product on the manufacturer's compliance with the inspection and recordkeeping requirements of this

Act and the Commission's rules with respect to such requirements.

EXPORTS

SEC. 18. This Act shall not apply to any consumer product if (1) it can be shown that such product is manufactured, sold, or held for sale for export from the United States (or that such product was imported for export), unless such consumer product is in fact distributed in commerce for use in the United States, and (2) such consumer product when distributed in commerce, or any container in which it is enclosed when so distributed, bears a stamp or label stating that such consumer product is intended for export; except that this Act shall apply to any consumer product manufactured for sale, offered for sale, or sold for shipment to any installation of the United States located outside of the United States.

PROHIBITED ACTS

SEC. 19. (a) It shall be unlawful for any person to—

(1)   manufacture for sale, offer for sale, distribute in commerce, or import into the United States any consumer product which is not in conformity with an applicable consumer product safety standard under this Act;

(2)   manufacture for sale, offer for sale, distribute in commerce, or import into the United States any consumer product which has been declared a banned hazardous product by a rule under this Act;

(3)   fail or refuse to permit access to or copying of records, *or fail or refuse to establish or maintain records,* or fail or refuse to make reports to provide information, or fail or refuse to permit entry or inspection, as required under this Act or rule thereunder;

(4)   fail to furnish information required by section 15(b);

(5)   fail to comply with an order issued under section 15(c) or (d) (relating to notification, to repair, replacement, and refund, *and to prohibited acts*);

(6)   fail to furnish a certificate required by section 14 or issue a false certificate if such person in the exercise of due care has reason to know that such certificate is false or misleading in any material respect; or to fail to comply with any rule under section 14(c) (relating to labeling);

(7)   fail to comply with any rule under section 9(d)(2) (relating to stockpiling); or

(8) *fail to comply with any rule under section 13 (relating to prior notice and description of new consumer products); or*

(9) *fail to comply with any rule under section 27(e) (relating to provision of performance and technical data).*

(b) Paragraphs (1) and (2) of subsection (a) of this section shall not apply to any person (1) who holds a certificate issued in accordance with section 14(a) to the effect that such consumer product conforms to all applicable consumer product safety rules, unless such person knows that such consumer product does not conform, or (2) who relies in good faith on the representation of the manufacturer or a distributor of such product that the product is not subject to an applicable product safety rule.

CIVIL PENALTIES

SEC. 20. (a)(1) Any person who knowingly violates section 19 of this Act shall be subject to a civil penalty not to exceed $2,000 for each such violation. Subject to paragraph (2), a violation of section 19(a) (1), (2), (4), (5), (6), (7), (8), or (9) shall constitute a separate offense with respect to each consumer product involved, except that the maximum civil penalty shall not exceed $500,000 for any related series of violations. A violation of section 19(a)(3) shall constitute a separate violation with respect to each failure or refusal to allow or perform an act required thereby; and, if such violation is a continuing one, each day of such violation shall constitute a separate offense, except that the maximum civil penalty shall not exceed $500,000 for any related series of violations.

(2) The second sentence of paragraph (1) of this subsection shall not apply to violations of paragraph (1) or (2) of section 19(a)—

(A) if the person who violated such paragraphs is not the manufacturer or private labeler or a distributor of the products involved, and

(B) if such person did not have either (i) actual knowledge that his distribution or sale of the product violated such paragraphs or (ii) notice from the Commission that such distribution or sale would be a violation of such paragraphs.

(b) Any civil penalty under this section may be compromised by the Commission. In determining the amount of such penalty or whether it should be remitted or mitigated and in what amount, the appropriateness of such penalty to the size of the business of the person charged and the gravity of the violation shall be considered. The amount of such penalty when finally determined or the amount agreed on compromise, may be deducted from any sums owing by the United States to the person charged.

(c) As used in the first sentence of subsection (a)(1) of this section. the term "knowingly" means (1) the having of actual knowledge, or (2) the presumed having of knowledge deemed to be possessed by a reasonable man who acts in the circumstances, including knowledge obtainable upon the exercise of due care to ascertain the truth of representations.

CRIMINAL PENALTIES

SEC. 21. (a) Any person who knowingly and willfully violates section 19 of this Act after having received notice of noncompliance from the Commission shall be fined not more than $50,000 or be imprisoned not more than one year, or both.

(b) Any individual director, officer, or agent of a corporation who knowingly and willfully authorizes, orders, or performs any of the acts or practices constituting in whole or in part a violation of section 19, and who has knowledge of notice of noncompliance received by the corporation from the Commission, shall be subject to penalties under this section without regard to any penalties to which that corporation may be subject under subsection (a).

INJUNCTIVE ENFORCEMENT AND SEIZURE

SEC. 22. (a) *The United States district courts shall have jurisdiction to take the following action:*

(1) *Restrain any violation of section 19.*

(2) *Restrain any person from manufacturing for sale, offering for sale, distributing in commerce, or importing into the United States a product in violation of an order in effect under section 15(d).*

(3) *Restrain any person from distributing in commerce a product which does not comply with a consumer product safety rule. Such actions may be brought by the Commission (without regard to section 27(b)(7)(A)) or by the Attorney General in any United States district court for a district wherein any act, omission, or transaction constituting the violation occurred, or in such court for the district wherein the defendant is found or transacts business. In any action under this section process may be served*

on a defendant in any other district in which the defendant resides or may be found.

(b) *Any consumer product—*

(1) *which fails to conform with an applicable consumer product safety rule, or*

(2) *the manufacture for sale, offering for sale, distribution in commerce, or the importation into the United States, of which has been prohibited by an order in effect under section 15(d),*

when introduced into or while in commerce or while held for sale after shipment in commerce shall be liable to be proceeded against on libel of information and condemned in any district court of the United States within the jurisdiction of which such consumer product is found. Proceedings in cases instituted under the authority of this subsection shall conform as nearly as possible to proceedings in rem in admiralty. Whenever such proceedings involving substantially similar consumer products are pending in courts of two or more judicial districts they shall be consolidated for trial by order of any such court upon application reasonably made by any party in interest upon notice to all other parties in interest.

SUITS FOR DAMAGES BY PERSONS INJURED

SEC. 23. (a) Any person who shall sustain injury by reason of any knowing (including willful) violation of a consumer product safety rule, or any other rule or order issued by the Commission may sue any person who knowingly (including willfully) violated any such rule or order in any district court of the United States in the district in which the defendant resides or is found or has an agent, subject to the provisions of section 1331 of title 28, United States Code as to the amount in controversy, shall recover damages sustained, and may, if the court determines it to be in the interest of justice, recover the costs of suit, including reasonable attorneys' fees (determined in accordance with section 10(e)(4)) and reasonable expert witnesses' fees.

(b) The remedies provided for in this section shall be in addition to and not in lieu of any other remedies provided by common law or under Federal or State law.

PRIVATE ENFORCEMENT OF PRODUCT SAFETY RULES AND OF SECTION 15 ORDERS

SEC. 24. Any interested person may bring an action in any United States district court for the district in which the defendant is found or transacts business to enforce a consumer product safety rule or an order under section 15, and to obtain appropriate injunctive relief. Not less than thirty days prior to the commencement of such action, such interested person shall give notice by registered mail to the Commission, to the Attorney General, and to the person against whom such action is directed. Such notice shall state the nature of the alleged violation of any such standard or order, the relief to be requested, and the court in which the action will be brought. No separate suit shall be brought under this section if at the time the suit is brought the same alleged violation is the subject of a pending civil or criminal action by the United States under this Act. In any action under this section the court may in the interest of justice award the costs of suit, including reasonable attorneys' fees (determined in accordance with section 10(e)(4)) and reasonable expert witnesses' fees.

EFFECT ON PRIVATE REMEDIES

SEC. 25. (a) Compliance with consumer product safety rules or other rules or orders under this Act shall not relieve any person from liability at common law or under State statutory law to any other person.

(b) The failure of the Commission to take any action or commence a proceeding with respect to the safety of a consumer product shall not be admissible in evidence in litigation at common law or under State statutory law relating to such consumer product.

(c) Subject to sections 6(a)(2) and 6(b) but notwithstanding section 6(a)(1), (1) any accident or investigation report made under this Act by an officer or employee of the Commission shall be made available to the public in a manner which will not identify any injured person or any person treating him, without the consent of the person so identified, and (2) all reports on research projects, demonstration projects, and other related activities shall be public information.

EFFECT ON STATE STANDARDS

SEC. 26. (a) Whenever a consumer product safety standard under this Act is in effect and applies to a risk of injury associated with a consumer product, no State or political subdivision of a State shall have any authority either to establish or to continue in effect any provision of a safety standard or regulation which prescribes any requirements as to the performance, composition, contents, design, finish, construction, packaging, or labeling of such products which are designed to deal with the same risk of injury associated with such consumer product, unless such requirements are identical to the requirements of the Federal standard.

(b) Subsection (a) of this section does not prevent the Federal Government or the government of any State or political subdivision of a State from establishing or continuing in effect a safety requirement applicable to a consumer product for its own use which requirement is designed to protect against a risk of injury associated with the product and which is not identical to the consumer product safety standard applicable to the product under this Act if the Federal, State, or political subdivision requirement provides a higher degree of protection from such risk of injury than the standard applicable under this Act.

(c) Upon application of a State or political subdivision of a State, the Commission may rule, after notice and opportunity for oral presentation of views, exempt from the provisions of subsection (a) (under such conditions as it may impose in the rule) any proposed safety standard or regulation which is described in such application and which is designed to protect against a risk of injury associated with a consumer product subject to a consumer product safety standard under this Act if the State or political subdivision standard or regulation—

(1) provides a significantly higher degree of protection from such risk of injury than the consumer product safety standard under this Act, and

(2) does not unduly burden interstate commerce.

In determining the burden, if any, of a State or political subdivision standard or regulation on interstate commerce, the Commission shall consider and make appropriate (as determined by the Commission in its discretion) findings on the technological and economic feasibility of complying with such standard or regulation, the cost of complying with such standard or regulation, the geographic distribution of the consumer product to which the standard or regulation would apply, the probability of other States or political subdivisions applying for an exemption under this subsection for a similar standard or regulation, and the need for a national, uniform standard under this Act for such consumer product.

ADDITIONAL FUNCTIONS OF COMMISSION

SEC. 27. (a) The Commission may, by one or more of its members or by such agents or agency as it may designate, conduct any hearing or other inquiry necessary or appropriate to its functions anywhere in the United States. A Commissioner who participates in such a hearing or other inquiry shall not be disqualified solely by reason of such participation from subsequently participating in a decision of the Commission in the same matter. The Commission shall publish notice of any proposed hearing in the Federal Register and shall afford a reasonable opportunity for interested persons to present relevant testimony and data.

(b) The Commission shall also have the power—

(1) to require, by special or general orders, any person to submit in writing such reports and answers to questions as the Commission may prescribe; and such submission shall be made within such reasonable period and under oath or otherwise as the Commission may determine;

(2) to administer oaths;

(3) to require by subpoena the attendance and testimony of witnesses and the production of all documentary evidence relating to the execution of its duties;

(4) in any proceeding or investigation to order testimony to be taken by deposition before any person who is designated by the Commission and has the power to administer oaths and, in such instances, to compel testimony and the production of evidence in the same manner as authorized under paragraph (3) of this subsection;

(5) to pay witnesses the same fees and mileage as are paid in like circumstances in the courts of the United States;

(6) to accept gifts and voluntary and uncompensated services, notwithstanding the provisions of section 3679 of the Revised Statutes (31 U.S.C. 665(b));

(7) to—

(A) initiate, prosecute, defend, or appeal (other than to the Supreme Court of the United States), through its own legal representative and in the name of the Commission, any civil action if the Commission makes a written request to the Attorney General for representation in such civil action and the Attorney General does not within the 45-day period beginning on the date such request was made notify the Commission in writing that the Attorney General will represent the Commission in such civil action, and

(B) initiate, prosecute, or appeal, through its own legal representative, with the concurrence of the Attorney General or through the Attorney General, any criminal action, for the purpose of enforcing the laws subject to its jurisdiction;

(8) to lease buildings or parts of buildings in the District of Columbia, without regard to the Act of March 3, 1877 (40 U.S.C. 34), for the use of the Commission; and

(9) to delegate any of its functions or powers, other than the power to issue subpoenas under paragraph (3), to any officer or employee of the Commission.

(c) Any United States district court within the jurisdiction of which any inquiry is carried on, may, upon petition by the Commission (subject to subsection (b)(7)) or by the Attorney General, in case of refusal to obey a subpoena or order of the Commission issued under subsection (b) of this section, issue an order requiring compliance therewith; and any failure to obey the order of the court may be punished by the court as a contempt thereof.

(d) No person shall be subject to civil liability to any person (other than the Commission or the United States) for disclosing information at the request of the Commission.

(e) The Commission may by rule require any manufacturer of consumer products to provide to the Commission such performance and technical data related to performance and safety as may be required to carry out the purposes of this Act, and to give such notification of such performance and technical data at the time of original purchase to prospective purchasers and to the first purchaser of such product for purposes other than resale, as it determines necessary to carry out the purposes of this Act.

(f) For purposes of carrying out this Act, the Commission may purchase any consumer product and it may require any manufacturer, distributor, or retailer of a consumer product to sell the product to the Commission at manufacturer's , distributor's, or retailer's cost.

(g) The Commission is authorized to enter into contracts with governmental entities, private organizations, or individuals for the conduct of activities authorized by this Act.

(h) The Commission may plan, construct, and operate a facility or facilities suitable for research, development, and testing of consumer products in order to carry out this Act.

(i) (1) Each recipient of assistance under this Act pursuant to grants or contracts entered into

under other than competitive bidding procedures shall keep such records as the Commission by rule shall prescribe, including records which fully disclose the amount and disposition by such recipient of the proceeds of such assistance, the total cost of the project undertaken in connection with which such assistance is given or used, and the amount of that portion of the cost of the project or undertaking supplied by other sources, and such other records as will facilitate an effective audit.

(2) The Commission and the Comptroller General of the United States, or their duly authorized representatives, shall have access for the purpose of audit and examination to any books, documents, papers, and records of the recipients that are pertinent to the grants or contracts entered into under this Act under other than competitive bidding procedures.

(j) The Commission shall prepare and submit to the President and the Congress on or before October 1 of each year a comprehensive report on the administration of this Act for the preceding fiscal year. Such report shall include—

(1) a thorough appraisal, including statistical analyses, estimates, and long-term projections, of the incidence of injury and effects to the population resulting from consumer products, with a breakdown, insofar as practicable, among the various sources of such injury;

(2) a list of consumer product safety rules prescribed or in effect during such year;

(3) an evaluation of the degree of observance of consumer product safety rules, including a list of enforcement actions, court decisions, and compromises of alleged violations, by location and company name;

(4) a summary of outstanding problems confronting the administration of this Act in order of priority;

(5) an analysis and evaluation of public and private consumer product safety research activities;

(6) a list, with a brief statement of the issues, of completed or pending judicial actions under this Act;

(7) the extent to which technical information was disseminated to the scientific and commercial communities and consumer information was made available to the public;

(8) the extent of cooperation between Commission officials and representatives of industry and other interested parties in the implementation of this Act, including a log or

summary of meetings held between Commission officials and representatives of industry and other interested parties;

(9)   an appraisal of significant actions of State and local governments relating to the responsibilities of the Commission; and

(10)   such recommendations for additional legislation as the Commission deems necessary to carry out the purposes of this Act.

(k)   (1) Whenever the Commission submits any budget estimate or request to the President or the Office of Management and Budget, it shall concurrently transmit a copy of that estimate or request to the Congress.

(2)   Whenever the Commission submits any legislative recommendations, or testimony, or comments on legislation to the President or the Office of Management and Budget, it shall concurrently transmit a copy thereof to the Congress. No officer or agency of the United States shall have any authority to require the Commission to submit its legislative recommendations, or testimony, or comments on legislation, to any officer or agency of the United States for approval, comments, or review, prior to the submission of such recommendations, testimony, or comments to the Congress.

(1)   (1) *Except as provided in paragraph (2)—*

(A)   the Commission shall transmit to the Committee on Commerce of the Senate and the Committee on Interstate and Foreign Commerce of the House of Representatives each consumer product safety rule proposed after the date of the enactment of this subsection and each regulation proposed by the Commission after such date under section 2 or 3 of the Federal Hazardous Substances Act, section 3 of the Poison Prevention Packaging Act of 1970, or section 4 of the Flammable Fabrics Act; and

(B)   no consumer product safety rule and no regulation under a section referred to in subparagraph (A) may be adopted by the Commission before the thirtieth day after the date the proposed rule or regulation upon which such rule or regulation was based was transmitted pursuant to subparagraph (A).

(2)   Paragraph (1) does not apply with respect to a regulation under section 2(q) of the Federal Hazardous Substances Act respecting a hazardous substance the distribution of which is found under paragraph (2) of such section to present an imminent hazard or a regulation under section 3(e) of such Act respecting a toy or

other article intended for use by children the distribution of which is found under paragraph (2) of such section to present an imminent hazard.

PRODUCT SAFETY ADVISORY COUNCIL

SEC. 28. (a) The Commission shall establish a Product Safety Advisory Council which it may consult before prescribing a consumer product safety rule or taking other action under this Act. The Council shall be appointed by the Commission and shall be composed of fifteen members, each of whom shall be qualified by training and experience in one or more of the fields applicable to the safety of products within the jurisdiction of the Commission. The Council shall be constituted as follows:

(1)   five members shall be selected from governmental agencies including Federal, State, and local governments;

(2)   five members shall be selected from consumer product industries including at least one representative of small business; and

(3)   five members shall be selected from among consumer organizations, community organizations, and recognized consumer leaders.

(b)   The Council shall meet at the call of the Commission, but not less often than four times during each calendar year.

(c)   The Council may propose consumer product safety rules to the Commission for its consideration and may function through subcommittees of its members. All proceedings of the Council shall be public, and a record of each proceeding shall be available for public inspection.

(d)   Members of the Council who are not officers or employees of the United States shall, while attending meetings or conferences of the Council or while otherwise engaged in the business of the Council, be entitled to receive compensation at a rate fixed by the Commission, not exceeding the daily equivalent of the annual rate of basic pay in effect for grade GS-18 on the General Schedule, including travel time, and while away from their homes or regular places of business they may be allowed travel expenses, including per diem in lieu of subsistence, as authorized by section 5703 of title 5, United States Code. Payments under this subsection shall not render members of the Council officers or employees of the United States for any purpose.

COOPERATION WITH STATES AND WITH OTHER FEDERAL AGENCIES

SEC. 29. (a) The Commission shall establish a program to promote Federal-State cooperation for the purposes of carrying out this Act. In implementing such program the Commission may—

(1) accept from any State or local authorities engaged in activities relating to health, safety, or consumer protection assistance in such functions as injury data collection, investigation, and educational programs, as well as other assistance in the administration and enforcement of this Act which such States or localities may be able and willing to provide and, if so agreed, may pay in advance or otherwise for the reasonable cost of such assistance, and

(2) commission any qualified officer or employee of any State or local agency as an officer of the Commission for the purpose of conducting examinations, investigations, and inspections.

(b) In determining whether such proposed State and local programs are appropriate in implementing the purposes of this Act, the Commission shall give favorable consideration to programs which establish separate State and local agencies to consolidate functions relating to product safety and other consumer protection activities.

(c) The Commission may obtain from any Federal department or agency such statistics, data, program reports, and other materials as it may deem necessary to carry out its functions under this Act. Each such department or agency may cooperate with the Commission and, to the extent permitted by law, furnish such materials to it. The Commission and the heads of other departments and agencies engaged in administering programs related to product safety shall, to the maximum extent practicable, cooperate and consult in order to insure fully coordinated efforts.

(d) The Commission shall, to the maximum extent practicable, utilize the resources and facilities of the National Bureau of Standards, on a reimbursable basis, to perform research and analyses related to risks of injury associated with consumer products (including fire and flammability risks), to develop test methods, to conduct studies and investigations, and to provide technical advice and assistance in connection with the functions of the Commission.

(e) *The Commission may provide to another Federal agency or a State or local agency or authority engaged in activities relating to health, safety, or consumer protection, copies of any accident or investigation report made under this Act by any officer, employee, or agent of the Commission only if (1) information which under section 6(a)(2) is to be considered confidential is not included in any copy of such report which is provided under this subsection; and (2) each Federal agency and State and local agency and authority which is to receive under this subsection a copy of such report provides assurances satisfactory to the Commission that the identity of any injured person and any person who treated an injured person will not, without the consent of the person identified, be included in—*

(A) *any copy of any such report, or*

(B) *any information contained in any such report which the agency or authority makes available to any member of the public. No Federal agency or State or local agency or authority may disclose to the public any information contained in a report received by the agency or authority under this subsection unless with respect to such information the Commission has complied with the applicable requirements of section 6(b).*

TRANSFERS OF FUNCTIONS

SEC. 30. (a) The functions of the Secretary of Health, Education, and Welfare under the Federal Hazardous Substances Act (15 U.S.C. 1261 et seq.) and the Poison Prevention Packaging Act of 1970 are transferred to the Commission. The functions of the Secretary of Health, Education, and Welfare under the *Federal Food, Drug, and Cosmetic Act (15 U.S.C. 301 et seq.)*, to the extent such functions relate to the administration and enforcement of the Poison Prevention Packaging Act of 1970, are transferred to the Commission.

(b) The functions of the Secretary of Health, Education, and Welfare, the Secretary of Commerce, and the Federal Trade Commission under the Flammable Fabrics Act (15 U.S.C. 1191 et seg.) are transferred to the Commission. The functions of the Federal Trade Commission under the Federal Trade Commission Act, to the extent such functions relate to the administration and enforcement of the Flammable Fabrics Act, are transferred to the Commission.

(c) The functions of the Secretary of Commerce and the Federal Trade Commission under the Act of August 2, 1956 (15 U.S.C. 1211) are transferred to the Commission.

(d) A risk of injury which is associated with a consumer product and which could be eliminated or reduced to a sufficient extent by action under the Federal Hazardous Substances Act, the Poison Prevention Packaging Act of 1970, or the Flammable Fabrics Act may be regulated under this Act only if the Commission by rule finds that it is in the public interest to regulate such risk of injury under this Act. Such a rule shall identify the risk of injury proposed to be regulated under this Act and shall be promulgated in accordance with section 553 of title 5, United States Code; except that the period to be provided by the Commission pursuant to subsection (c) of such section for the submission of data, views, and arguments respecting the rule shall not exceed thirty days from the date of publication pursuant to subsection (b) of such section of a notice respecting the rule.

(e) (1) (A) All personnel, property, records, obligations, and commitments, which are used primarily with respect to any function transferred under the provisions of subsections (a), (b), and (c) of this section shall be transferred to the Commission, except those associated with fire and flammability research in the National Bureau of Standards. The transfer of personnel pursuant to this paragraph shall be without reduction in classification or compensation for one year after such transfer, except that the Chairman of the Commission shall have full authority to assign personnel during such one-year period in order to efficiently carry out functions transferred to the Commission under this section.

(B) Any commissioned officer of the Public Health Service who upon the day before the effective date of this section, is serving as such officer primarily in the performance of functions transferred by this Act to the Commission, may, if such officer so elects, acquire competitive status and be transferred to a competitive position in the Commission subject to subparagraph (A) of this paragraph, under the terms prescribed in paragraphs (3) through (8)(A) of section 15(b) of the Clean Air Amendments of 1970 (84 Stat. 1676; 42 U.S.C. 215 nt).

(2) All orders, determinations, rules, regulations, permits, contracts, certificates, licenses, and privileges (A) which have been issued, made, granted, or allowed to become effective in

the exercise of functions which are transferred under this section by any department or agency, any functions of which are transferred by this section, and (B) which are in effect at the time this section takes effect, shall continue in effect according to their terms until modified, terminated, superseded, set aside, or repealed by the Commission, by any court of competent jurisdiction, or by operation of law.

(3) The provisions of this section shall not affect any proceeedings pending at the time this section takes effect before any department or agency, functions of which are transferred by this section; except that such proceedings, to the extent that they relate to functions so transferred, shall be continued before the Commission. Order shall be issued in such proceedings, appeals shall be taken therefrom, and payments shall be made pursuant to such orders, as if this section had not been enacted; and orders issued in any such proceedings shall continue in effect until modified, terminated, superseded, or repealed by the Commission, by a court of competent jurisdiction, or by operation of law.

(4) The provisions of this section shall not affect suits commenced prior to the date this section takes effect and in all such suits proceedings shall be had, appeals taken, and judgments rendered, in the same manner and effect as if this section had not been enacted; except that if before the date on which this section takes effect, any department or agency (or officer thereof in his official capacity) is a party to a suit involving functions transferred to the Commission, then such suit shall be continued by the Commission. No cause of action, and no suit, action, or other proceeding, by or against any department or agency (or officer thereof in his official capacity) functions of which are transferred by this section, shall abate by reason of the enactment of this section. Causes of actions, suits, actions, or other proceedings may be asserted by or against the United States or the Commission as may be appropriate and, in any litigation pending when this section takes effect, the court may at any time, on its own notion or that of any party, enter an order which will give effect to the provisions of this paragraph.

(f) For purposes of this section, (1) the term "function" includes power and duty, and (2) the transfer of a function, under any provision of law, of an agency or the head of a department shall also be a transfer of all functions under such law which are exercised by any office or officer of such agency or department.

LIMITATION ON JURISDICTION

SEC. 31. The Commission shall have no authority under this Act to regulate any risk of injury associated with a consumer product if such risk could be eliminated or reduced to a sufficient extent by actions taken under the Occupational Safety and Health Act of 1970; the Atomic Energy Act of 1954; or the Clean Air Act. The Commission shall have no authority under this Act to regulate any risk of injury associated with electronic product radiation emitted from an electronic product (as such terms are defined by sections 355 (1) and (2) of the Public Health Service Act) if such risk of injury may be subjected to regulation under subpart 3 of part F of title III of the Public Health Service Act.

AUTHORIZATION OF APPROPRIATIONS

SEC. 32. (a) There are authorized to be appropriated for the purposes of carrying out the provisions of this Act (other than the provisions of section 27(h) which authorizes the planning and construction of research, development, and testing facilities) and for the purpose of carrying out the functions, powers and duties transferred to the Commission under section 30, not to exceed—

(1) $51,000,000 for the fiscal year ending June 30, 1977;

(2) $14,000,000 for the period beginning July 1, 1976, and ending September 30, 1976;

(3) $60,000,000 for the fiscal year ending September 30, 1977; and

(4) $68,000,000 for the fiscal year ending September 30, 1978.

(b) (1) There are authorized to be appropriated such sums as may be necessary for the planning and construction of research, development and testing facilities described in section 27(h); except that no appropriation shall be made for any such planning or construction involving an expenditure in excess of $100,000 if such planning or construction has not been approved by resolutions adopted in substantially the same form by the Committee on Interstate and Foreign Commerce of the Senate. For the purpose of securing consideration of such

approval the Commission shall transmit to Congress a prospectus of the proposed facility including (but not limited to)—

(A) a brief description of the facility to be planned or constructed;

(B) the location of the facility, and an estimate of the maximum cost of the facility;

(C) a statement of those agencies, private and public, which will use such facility, together with the contribution to be made by each such agency toward the cost of such facility; and

(D) a statement of justification of the need for such facility.

(2) The estimated maximum cost of any facility approved under this subsection as set forth in the prospectus may be increased by the amount equal to the percentage increase, if any, as determined by the Commission, in construction costs, from the date of the transmittal of such prospectus to Congress, but in no event shall the increase authorized by this paragraph exceed 10 per centum of such estimated maximum cost.

(c) *No funds appropriated under subsection (a) may be used to pay any claim described in section 4(i) whether pursuant to a judgment of a court or under any award, compromise, or settlement of such claim made under section 2672 of title 28, United States Code, or under any other provision of law.*

SEPARABILITY

SEC. 33. If any provision of this Act, or the application of such provision to any person or circumstance, shall be held invalid, the remainder of this Act, or the application of such provisions to persons or circumstances other than those as to which it is held invalid, shall not be affected thereby.

EFFECTIVE DATE

SEC. 34. This Act shall take effect on the sixtieth day following the date of its enactment, except—

(1) sections 4 and 32 shall take effect on the date of enactment of this Act, and

(2) section 30 shall take effect on the later of (A) 150 days after the date of enactment of this Act, or (B) the date on which at least three members of the Commission first take office.

APPENDIX 3:
Selected Articles from the Uniform Commercial Code

2-311. Options and Cooperation Respecting Performance.

(1) An agreement for sale which is otherwise sufficiently definite (subsection (3) of Section 2-204) to be a contract is not made invalid by the fact that it leaves particulars of performance to be specified by one of the parties. Any such specification must be made in good faith and within limits set by commercial reasonableness.

(2) Unless otherwise agreed specifications relating to assortment of the goods are at the buyer's option and except as otherwise provided in subsections (1) (c) and (3) of Section 2-319 specifications or arrangements relating to shipment are at the seller's option.

(3) Where such specification would materially affect the other party's performance but is not seasonally made or where one party's cooperation is necessary to the agreed performance of the other but is not seasonably forthcoming, the other party in addition to all other remedies
  (a) is excused for any resulting delay in his own performance; and
  (b) may also either proceed to perform in any reasonable manner or after the time for a material part of his own performance treat the failure to specify or to cooperate as a breach by failure to deliver or accept the goods.

2-312. Warranty of Title and Against Infringement; Buyer's Obligation Against Infringement.

(1) Subject to subsection (2) there is in a contract for sale a warranty by the seller that
  (a) the title conveyed shall be good, and its transfer rightful; and
  (b) the goods shall be delivered free from any security interest or other lien or encumbrance of which the buyer at the time of contracting has no knowledge.

(2) A warranty under subsection (1) will be excluded or modified only by specific language or by circumstances which give the buyer reason to know that the person selling does not claim title in himself or that he is purporting to sell only such right or title as he or a third person may have.

(3) Unless otherwise agreed a seller who is a merchant regularly dealing in goods of the kind warrants that the goods shall be delivered free of the

rightful claim of any third person by way of infringement or the like but a buyer who furnishes specifications to the seller must hold the seller harmless against any such claim which arises out of compliance with the specifications.

2-313. Express Warranties by Affirmation, Promise, Description, Sample.

(1) Express warranties by the seller are created as follows:

   (a) Any affirmation of fact or promise made by the seller to the buyer which relates to the goods and becomes part of the basis of the bargain creates an express warranty that the goods shall conform to the affirmation or promise.

   (b) Any description of the goods which is made part of the basis of the bargain creates an express warranty that the goods shall conform to the description.

   (c) Any sample or model which is made part of the basis of the bargain creates an express warranty that the whole of the goods shall conform to the sample or model.

(2) It is not necessary to the creation of an express warranty that the seller use formal words such as "warrant" or "guarantee" or that he have a specific intention to make a warranty, but an affirmation merely of the value of the goods or a statement purporting to be merely the seller's opinion or commendation of the goods does not create a warranty.

2-314. Implied Warranty; Merchantability; Usage of Trade.

(1) Unless excluded or modified (Section 2-316), a warranty that the goods shall be merchantable is implied in a contract for their sale if the seller is a merchant with respect to goods of that kind. Under this section the serving for value of food or drink to be consumed either on the premises or elsewhere is a sale.

(2) Goods to be merchantable must be at least such as

   (a) pass without objection in the trade under the contract description; and

   (b) in the case of fungible goods, are of fair average quality within the description; and

   (c) are fit for the ordinary purposes for which such goods are used; and

   (d) run, within the variations permitted by the agreement, of even kind, quality and quantity within each unit and among all units involved; and

   (e) are adequately contained, packaged, and labeled as the agreement may require, and

   (f) conform to the promises or affirmations of fact made on the container or label if any.

(3) Unless excluded or modified (Section 2-316) other implied warranties may arise from course of dealing or usage of trade.

2-315. Implied Warranty: Fitness for Particular Purpose.

Where the seller at the time of contracting has reason to know any particular purpose for which the goods are required and that the buyer is relying on the seller's skill or judgment to select or furnish suitable goods, there is unless excluded or modified under the next section an implied warranty that the goods shall be fit for such purpose.

2-316.   Exclusion or Modification of Warranties.

(1) Words or conduct relevant to the creation of an express warranty and words or conduct tending to negate or limit warranty shall be construed wherever reasonable as consistent with each other; but subject to the provisions of this Article on parol or extrinsic evidence (Section 2-202) negation or limitation is inoperative to the extent that such construction is unreasonable.

(2) Subject to subsection (3), to exclude or modify the implied warranty of merchantability or any part of it the language must mention merchantability and in case of a writing must be conspicuous, and to exclude or modify any implied warranty of fitness the exclusion must be by a writing and conspicuous. Language to exclude all implied warranties of fitness is sufficient if it states, for example, that "There are no warranties which extend beyond the description on the face hereof."

(3) Notwithstanding subsection (2)

    (a) unless the circumstances indicate otherwise, all implied warranties are excluded by expressions like "as is", "with all faults" or other language which in common understanding calls the buyer's attention to the exclusion of warranties and makes plain that there is no implied warranty; and

    (b) when the buyer before entering into the contract has examined the goods or the sample or model as fully as he desired or has refused to examine the goods there is no implied warranty with regard to defects which an examination ought in the circumstances to have revealed to him; and

    (c) an implied warranty can also be excluded or modified by course of dealing or course of performance or usage of trade.

(4) Remedies for breach of warranty can be limited in accordance with the provisions of this Article on liquidation or limitation of damages and on contractual modification of remedy (Sections 2-718 and 2-719).

2-317.   Cumulation and Conflict of Warranties Express or Implied.

Warranties whether express or implied shall be construed as consistent with each other and as cumulative, but if such construction is unreasonable, the intention of the parties shall determine which warranty is dominant. In ascertaining that intention the following rules apply:

    (a) Exact or technical specifications displace an inconsistent sample or model or general language of description.

    (b) A sample from an existing bulk displaces inconsistent general language of description.

    (c) Express warranties displace inconsistent inplied warranties other than an implied warranty of fitness for a particular purpose.

2-318.   Third Party Beneficiaries of Warranties Express or Implied.

A seller's warranty whether express or implied extends to any natural person who is in the family or household of his buyer or who is a guest in his home if it is reasonable to expect that such person may use, consume or be affected by the goods and who is injured in person by breach of the warranty. A seller may not exclude or limit the operation of this section.

Index